FIELD NOTES ON SCIENCE & NATURE

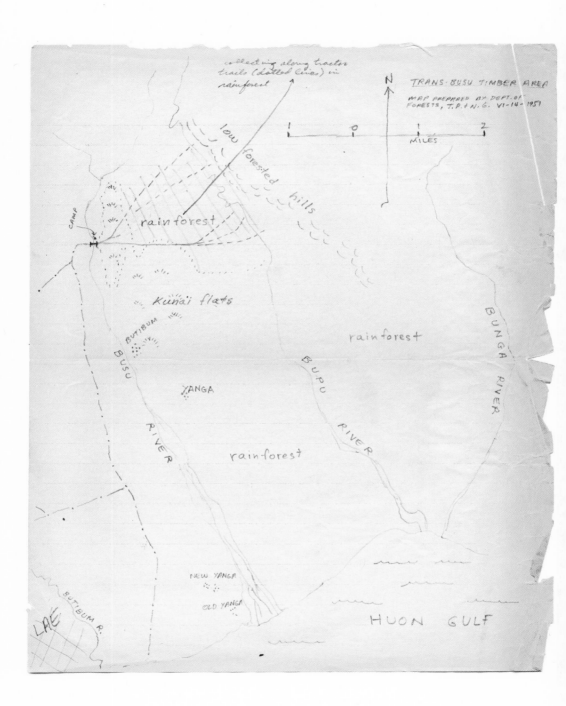

collecting along tractor
trails (dotted lines) in
rainforest

N

TRANS-BUSU TIMBER AREA

MAP PREPARED BY DEPT. OF
FORESTS, T.P. & N.G. VI-14-1957

1   O   1   2

MILES

low forested hills

rainforest

CAMP

Kunai flats

BUTIBUM

BUSU

RIVER

YANGA

rainforest

BUPU

RIVER

BUNGA RIVER

rainforest

LAE

BUTIBUM R.

NEW YANGA

OLD YANGA

HUON GULF

FIELD NOTES
on Science & Nature

EDITED BY

MICHAEL R. CANFIELD

HARVARD UNIVERSITY PRESS

Cambridge, Massachusetts | *London, England*

2011

Chapter 5 was prepared by Anna K. Behrensmeyer as part of official duties as an employee of the Federal Government/Smithsonian Institution. The copyright in *Field Notes on Science & Nature* does not extend to this chapter.

Many of the designations used by manufacturers and sellers to distinguish their products are claimed as trademarks. Where those designations appear in this book and Harvard University Press was aware of a trademark claim, then the designations have been printed in initial capital letters.

Library of Congress Cataloging-in-Publication Data

Field notes on science and nature / edited by Michael R. Canfield.
 p. cm.
 Includes bibliographical references and index.
 ISBN 978-0-674-05757-9 (alk. paper)
 1. Biology—Fieldwork. 2. Natural history—Fieldwork. 3. Note-taking.
I. Canfield, Michael R.
 QH318.5.F54 2011
 570.72—dc22 2011005780

FRONTISPIECE: A hand-drawn map of collecting localities in the Trans-Busu Timber Reserve near Lae, Morobe Province, Papua New Guinea, which was inserted in the back of a field notebook from E. O. Wilson's collecting expedition to the South Pacific from 1954 to 1955.

To Jen

CONTENTS

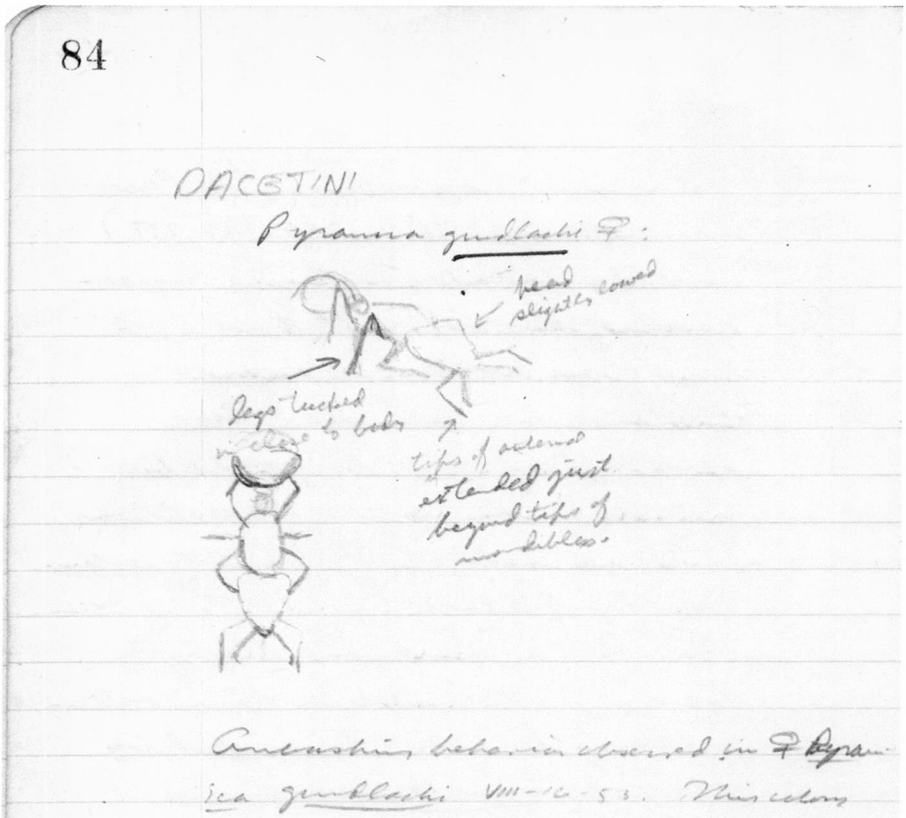
84

DACETINI

Pyramica gundlachi ♀:

head slightly lowered

legs tucked close to body

tips of antennae extended just beyond tips of mandibles.

Ambushing behavior observed in ♀ Pyramica gundlachi VIII-16-53. This colony has gone without food for a month and the brood has been totally consumed. The ♀ foraged with the workers when a new batch of collembolans were put in. The ♀ bumped into a collembolan (entomobryid) slightly smaller than she, touching its antennae with hers. She immediately backed away, stepped from right to left a couple of times hesitatingly, then advanced

Sketches and behavioral notes on *Pyramica gundlachi*, an ant species distributed throughout the Caribbean. August 16, 1953, Edward O. Wilson.

FOREWORD

EDWARD O. WILSON

THE SECOND HALF OF THE TWENTIETH CENTURY witnessed the rise of molecular and cellular biology, one of the greatest achievements in the history of science. The study of living things at the molecular level established what may be fairly called the First Law of Biology, that all the entities and processes of life are obedient to the laws of physics and chemistry. This research succeeded in part because it focused on several dozen species of "model organisms" to explore particular basic problems, for example, the colon bacterium *E. coli* for molecular genetics, the roundworm *C. elegans* for the molecular underpinnings of nerve cell development, and honeybees for the molecular basis of advanced social organization. Molecular biologists remained relatively unconcerned about the higher levels of biological organization, from organism to population, ecosystem, and society. They were also nonhistorical. As a consequence, relatively little attention was paid in the twentieth century to the Second Law of Biology, that all entities and processes of life were created by evolution through natural selection.

Biologists in the present century are beginning to shift toward more of a balance in attention to these two great divisions. Even at the level of the molecule and cell, a trend toward synthesis—how the entities and processes fit together to make a cell, then an organism—is rapidly growing stronger. Biology as a whole is turning to the comparison of species. The new goal is to understand the full diversity of life from molecule to ecosystem. It comes not a moment too soon. Humanity desperately needs a more extensive and integrated biology—for personal and public

52

Bird notes - New Caledonia

1. Mt. Mou, near summit, ca. 1200 meters XII-12-54. In true mossy forest, a parrot about the size of a starling; seen in a sitting position only and not flying, all green, except for top of head which was bright red, a little reddish around eyes, a faint yellow overtone on breast. Sat on a branch squawking at me about 15 feet away.

2. Mt. Mou, in bracken scrub, just above Pentecost residence, on south slope, at about 300 m. XII-12-54 Swallow-like birds, at least the wings were swallow-like, and they soared and dipped like swallows, black heads and tails, white bellies, grey wings and backs. A pair dove at my constantly at a certain point on the track on my way up early in the morning, and on my down in midafternoon. Brushed my hair many times, got in a couple of pecks also. Very aggressive & brave. (Fledglings, with short tail feathers, found in middle of trail XII-27-1954; parents dived frantically).

3. Fledgling found on ITO grounds, Anse Vata, Noumea, XII-23-54.

black eyes, yellow lit

life-size

Adults nearby; nest not found. Most of body, except for yellowish areas, brownish grey.

← yellow-tinged belly (although this was not noticed in parents)

with yellow,

Along with copious notes on ants, my notebooks contain natural history observations such as these notes on birds from New Caledonia, taken in July 1954.

health, support of biotechnology, resource management, conservation, and, not least, a more complete and wiser understanding of our
own species.

The wellspring of the new biology is scientific natural history.
That statement is not literary license. It is an axiom. Earth remains
a little-known planet. A large majority of its species (probably over
90 percent when microorganisms are included) are still unknown.
Of the approximately two million species that have been described
and given a scientific name, fewer than 10 percent have been studied
in depth. Many of the world's ecosystems have been examined only
cursorily, if at all. How, it should be asked, on the basis of so little
knowledge, are we ever going to know the living world, let alone
manage, preserve, and make full use of it?

Scientific naturalists are blessed. As researchers, everything
they touch turns to gold, because the living world is so little known.
For every kind of organism or ecosystem they choose to study, every
datum they collect is likely to be useful. Each morning they awaken
in the field they know there is a chance of a major discovery. While
no measure has yet been attempted, I feel certain that the rate of discoveries acquired per person-hour is an order of magnitude greater
in field studies than in the laboratory. That has been my experience
over many years. My personal record was set during a week I spent
studying ants with Bert Hölldobler at the field station of the Organization of Tropical Studies at La Selva, Costa Rica. We went joyously
from one previously unstudied ant species to another. From our rapidly filling notebooks and further work back home we published five
articles in peer-reviewed journals.

Naturalists know that what they see on a field trip, whether a
day's excursion or a year's residence in a camp or research station,
is only a tiny part of what exists around them. They also know that
whether they go to the Amazon or just to a city park near their home,
they can find biological novelty. Such is especially the case if they
choose a mostly unstudied group of species for observation—which
means 95 percent or more in the Amazon, and perhaps 60–70 percent in a city park.

The greatest adventure of my life was to climb to the summit of the mid-Sarawaget Range in Papua New Guinea, not just

because in 1955 I was the first nonnative to do so, but because I was able to search for ants across the pristine altitudinal forest zones. It is this sense of exploring a rich and mostly unknown world that gives natural history its primal stimulus. I felt no less of it while studying ants in the West Indies during my late seventies than I did in my early twenties.

If there is a heaven, and I am allowed entrance, I will ask for no more than an endless living world to walk through and explore. I will carry with me an inexhaustible supply of notebooks, from which I can send back reports to the more sedentary spirits (mostly molecular and cell biologists). Along the way I would expect to meet kindred spirits, among whom would be the authors of the essays in this book.

This notebook page documents collections of *Pheidole* ants, an arboreal earthworm found in the inflated internode of a rainforest shrub, as well as *Aneuretus* ants, the elusive group I sought in Sri Lanka in 1955.

FIELD NOTES ON SCIENCE & NATURE

Introduction

MICHAEL R. CANFIELD

THOSE WHO STUDY NATURE are bound by a shared curiosity and common traditions. Whether tracking gorillas in the Congo or terns above
the Arctic Circle, those who take to the field seek information on how
organisms live and behave, how they interact, and how the world has
been shaped by the forces of nature. This work is rich not only because
of the immeasurable diversity of life, but also because of the human
experience that inevitably arises with the study, adventure, and once-in-
a-lifetime sightings that take place in the field. Along with these intellectual and aesthetic ties, field scientists share time-honored traditions of
inquiry: careful observation, patient and arduous experimentation, and
persistence in the face of monsoons, parasites, and insults from snakes
and urticating plants. Field scientists also have a common set of tools
that include binoculars and hand lenses, field guides, good footwear,
and the most fundamental and simple of all field equipment: paper and
pencil. These final implements are perhaps the most important, and
are required for continuing the tradition of recording the science and
narrative of the field in notebooks and journals. Authors in this book
have varying opinions about the usefulness of modern record-keeping
technologies in the field, but nearly all would agree that paper and pencil
remain the standard because of their simplicity and reliability.

Meticulous record keeping is at the heart of good science, and this is
especially true for field scientists and naturalists. However, the status of
field record keeping has come into question in the recent age of technological proliferation, and the first principles of field recording are rarely

taught. Little guidance exists to help individuals develop this foundational skill, except perhaps for the ample and often accessible examples of notable nineteenth- and twentieth-century scientists. A brief search in almost any bookstore, new or used, will turn up the field records of the patron saint of field workers, Charles Darwin, published in *The Voyage of the* Beagle. This expansive field narrative has been reprinted under a variety of titles since it first appeared in 1839, though here I'll refer to it as the *Voyage*.[1] It can be daunting to hold up one's own records to Darwin's accounts, though to do so is an exercise that is both inspiring and misleading.

In the *Voyage*, Darwin provides an account, starting on October 8, 1835, in the Galápagos Islands, that describes his observations of the birds and reptiles of the archipelago, including the marine iguana, *Amblyrhynchus cristatus*:

> It is easy to drive these lizards down to any little point overhanging the sea, where they will sooner allow a person to catch hold of their tail than jump into the water. They do not seem to have any notion of biting; but when much frightened they squirt a drop of fluid from each nostril. One day I carried one to a deep pool left by the retiring tide, and threw it in several times as far as I was able. It invariably returned in a direct line to the spot where I stood . . . I several times caught this same lizard, by driving it down to a point, and though possessed of such perfect powers of diving and swimming, nothing would induce it to enter the water; and as often as I threw it in, it returned in the manner above described. Perhaps this singular piece of apparent stupidity may be accounted for by the circumstance, that this reptile has no enemy whatever on shore, whereas at sea it must often fall a prey to the numerous sharks.[2]

A close examination of this passage reveals the *Voyage* not as an example of notes, but rather a travel narrative Darwin honed and polished from the actual field notes he kept in his zoological notebooks and diary while aboard the HMS *Beagle*.[3] It is a relief that we should not necessarily expect our notebooks to read like a passage from the *Voyage*, with Darwin's artful, frank, and probing language. However, a view of the actual passage in his zoological notebooks that led to this description suggests that Darwin kept copious and detailed notes in the field that allowed him to produce such a complete and fulfilling narrative of his work:

Yet it is remarkable, that when shuffling over the tidal rocks it is scarcely possible to drive them into the water. From this reason, it is easy to
catch them by the tail, after driving them on a point.— They have no
idea of biting, & only sometimes when frightened squirt a drop of fluid
from each nostril.— Having seized a large one by the tail, I threw him
it several times into a good distance into a deep pool left by the retiring tide.— Invariably the Lizard it returned to the in the same direction from which it was thrown to the spot I stood. Its motion was rapid,
swimming at the bottom of the water & occasionally helping itself by its
feet on the stones.— As soon as it was near the margin, it either tried to
conceal itself in the seaweed or entered some hole or crack. As soon as
it thought the danger was over it crawled out on the dry stones, & again
would sooner be caught than voluntarily enter the water.— What can
be the reason of this? are its habitual enemies sharks or other marine
animals?[4]

Undoubtedly, much can be learned about field notes and much else
from studying Darwin. However, the process of fieldwork has drastically
changed since Darwin boarded the *Beagle* in 1831.

When I went to the field as a graduate student, I spent many long
nights chasing moths. The following mornings I worked to record my
observations and experiments in notebooks. Like many others who study
in the field, my work combined elements of both science and natural
history. I had read Darwin's *Voyage* and seen fragments of Henry Walter
Bates's journals, but when I considered the scratches and scribbles in
my field notebooks, they seemed inadequate. Discouraged, I began looking for models to analyze as I worked to hone my ability to create useful
and organized documentation of my fieldwork. These were surprisingly
elusive. It was not until late on the night before an ant-collecting expedition, as I lay awake on a foldout couch in Roger Kitching's study, that I
found a different model for how a modern naturalist and field scientist
keeps his notes. In the shadow of his trophy collection of field guides,
I perused the bank of field notebooks that he had invited me to consider that afternoon. I stayed up late into the night paging through the
detailed notes of biological adventures and sketches of specimens that
were contained in his field journals. As I finally drifted off to sleep in the
muggy Australian evening, I realized that viewing the actual field notes

the tidal rocks it is scarcely possible
to drive them into the water. From
this reason. it is easy to catch them
by the tail. after driving them on a
point. — They have no idea of biting, &
only sometimes. when frightened squirt a drop
of fluid from each nostril. — Having
seized a large one by the tail. I threw
him it several times into a good distance
into a deep pool. left by the retiring
tide. — Invariably the lizard it returned
to the in the same direction. from which
it was thrown to the spot. where I stood.
Its motion was rapid. swimming at the
bottom of the water & occasionally helping
itself by its feet on the stones. — As
soon as it was near the margin. it
either tried to conceal itself in the seaweed or entered some hole or crack. As
soon as it thought the danger. was
over, crawled out on the dry stones. &
again would sooner be caught than voluntarily enter the water. — What can be
the reason of this? are its habitual enemies

of another scientist gave me new ideas about how I would construct my own.

We left for the bush early the next day—too soon, for I wished that I could return and again page through those vignettes, anecdotes, and sketches. This book is the result of my pursuit of other examples of field notes from working scientists and naturalists. Thirteen who represent diverse disciplines have contributed to this volume. These authors have been asked to provide excerpts from their field notes along with their perspectives on how field notes could or should be kept, problems and solutions they have encountered, and lore from the field. The following chapters provide examples and advice from eminent living field scientists and naturalists on how to keep field notes and on the possible ways to construct records across disciplines. This book is not a methods manual but rather offers a glimpse into the lives of some well-known naturalists and their diverse ways of recording nature. But before delving in, let's briefly consider the scope of the topic. So what are "field notes"? For that matter, what is "the field"?

Those who head to the field have their own understanding of its location and character. To some it evokes somewhere remote, to others it is close to home. This usage of "the field" first appears in a letter written to Gilbert White in response to his publication of one of the most important books in natural history, *The Natural History and Antiquities of Selborne* (1789), which describes the nature of his home parish in southern England.[5] Despite its eighteenth-century beginning, "the field" only came into common usage toward the end of the nineteenth century after such scientists as Darwin, Henry Walter Bates, and Alfred Russel Wallace took to the field to collect specimens and understand the principles of nature. The scope of field science widened at the beginning of the twentieth century, and so solidified the field as a place for study away from home or lab. Since this place mixes scientific pursuits with exposure to new terrain, languages, and peoples, and has an inseparable aspect of adventure, a narrative of the field has also emerged. The

(*opposite*) **Excerpt from Charles Darwin's zoological notebook concerning the behavior of the marine iguana (*Amblyrhynchus cristatus*) in the Galápagos Islands. The last passage, continued on the following notebook page, reads: "What can be the reason of this? are its habitual enemies sharks or other marine animals?"** Reproduced by the kind permission of the Syndics of the Cambridge University Library, manuscript DAR.31.2.

field has no geographical or physical bounds, but is defined by those who go there to investigate, study, or commune with nature. To a young naturalist, the field may come to life with unbounded imagination in an undeveloped lot. Others may find the field after long hours in a dugout canoe, dangerous river crossings, or battles with tropical diseases. Given the diversity of people and concepts of the field, there is no rigid formula for documenting the discoveries and adventures that happen there. However, a genre of record keeping—field notes—exists as a critical component of the study and experience of the field.

The emergent tradition of field notes is evident in the nascent stages of natural science. The history of field notes has not been written, nor will it be written here. However, the notes of some historical naturalists are available either as published accounts or as online archives, and these documents reveal something of the antecedents of modern field notes. For example, Carl Linnaeus, in addition to devising the classification system that we now use to describe all living things, kept a careful journal on his field excursions to Lapland and other parts of Sweden. The copious notes and sketches contained in Linnaeus's Lapland journal show his attention to detail and dedication to creating thorough records while in the field.[6]

Linnaeus himself only spent part of his time in the field and relied on the findings of the early naturalist-explorers who combed the globe for collections and new insights on nature. One of the earliest and most colorful of these was certainly the pirate-naturalist William Dampier. At the end of the seventeenth century, Dampier traveled with a pirate band that ransacked villages and plundered unlucky merchant ships.[7] In his free time, he observed birds and animals, kept detailed meteorological records, and eventually circumnavigated the globe a record three times. While his compatriots spent their evenings sharpening sabers and drinking rum, Dampier wrote copious field notes that he eventually published in *A New Voyage Round the World* and several other texts. Dampier recounts his dedication to his records while in Central America in 1681:

> Foreseeing a necessity of wading through Rivers frequently in our Landmarch, I took care before I left the Ship to provide my self a large Joint of Bambo, which I stopt at both ends, closing it with Wax, so as to keep out any Water. In this I preserved my Journal and other Writings from being wet, tho' I was often forced to swim.[8]

A page of Linnaeus's Lapland journal entries from June, 1732, concerning his observations of mosses, a lichen, a fly; also, detailed descriptions of several plants and a sketch of Andromeda (facsimile in *Iter Lapponicum: Lappländska resan 1732. Vol. III*). Used by permission from the Linnean Society of London.

Dampier's original journals have since been lost, but his example should cause contemporary naturalists to pause before grumbling about any environmental trials that prevent good record keeping. The notes Dampier took, and the publications that resulted, were important for both the meteorological data they contained and for their natural history. Indeed, Darwin makes frequent references to "old Dampier" in his notebooks and *Voyage*.[9]

The work of Dampier was also carefully noted by Captain James Cook, who led one of the most important early expeditions from 1768–1771 aboard the HMS *Endeavour*. The naturalist Joseph Banks was recruited to document the natural findings during that journey, which he did in his own careful field notes as well as with the help of several artists. On July 26, 1770, Banks recounts:

> In botanizing to day I had the good fortune to take an animal of the Opossum (*Didelphis*) tribe: it was a female and with it I took two young ones. It was not unlike that remarkable one which De Bufon has decribd by the name of Phalanger as an American animal; it was however not the same for De Buffon is certainly wrong in asserting that this tribe is peculiar to America; and in all probability, as Pallas has said in his *Zoologia*, the Phalanger itself is a native of the East Indies, as my animal and that agree in the extrordinary conformation of their feet in which particular they differ from all the others.[10]

Other eminent nineteenth-century scientists kept careful field notes, and many such as Richard Spruce, Alfred Russel Wallace, and Henry Walter Bates published accounts of their journals.[11] *Field Notes on Science & Nature* picks up the tradition of field recording in the twentieth century, when field workers benefited from new access to remote locations and an array of increasingly quantified approaches. Elements of the tradition of naturalists' journals are still relevant to those who study in the field, and new approaches necessitate a reevaluation of how information should be captured while away from home or lab. Even as field notes themselves have a rich history, so too do the ways in which field note methodologies have been communicated.

Ever since the concept of the field became rooted at the time of Gilbert White, there have been attempts to communicate note-taking methodology. One of the earliest was Daines Barrington's *The Naturalist's Journal*.[12]

Initially published in 1767, Barrington's notebook lays out a template spreadsheet of daily weather conditions and observations on plants and animals to be filled in by the owner. Even White adopted this system after Barrington sent him a copy of his *Journal,* and he used it consistently until his death.

Joseph Banks's notes from July 26, 1770, on what he called the Endeavours River (now known as the Endeavour River) in the northern part of Queensland, Australia. The HMS *Endeavour* had been damaged on a coral reef just outside the river six weeks before, and the ship and crew had remained in the inlet to make repairs. Used with permission from the Mitchell Library, State Library of New South Wales.

17.

Year Selborne.	Place. Soil.	Barom.º	Therm.º	Wind.	Inches of Rain or Sn. Size of Hail-ñ.	Weather.	Trees first in leaf.—Fungi first appear.	Plants first in flower: Mosses vegetate.	Birds and Insects first appear, or disappear.	Observations with regard to fish, and other animals.	Miscellaneous Observations, and Memorandums,
April 18.	Sunday. 8 / 12 / 4 / 8	29¼ ⁷⁄₁₀	51	S. / W.		soft rain, rain. / sun. / bright.			Luscinia.		Ground very wet. Nightingale sings.
19.	Monday. 8 / 12 / 4 / 8	29¼ ⁷⁄₁₀	44 / 48	W. / SW. / S.		frost, sun, clouds. / soft rain. / cloudy & rain.	mercurialis perennis. Pellitoria bolaghag. pericularis Cochleaica		Atricapilla.		Blackcap sings. The sedge-bird, a delicate polyglott.
20.	Tuesday. 8 / 12 / 4 / 8	29¼ ¹⁰⁄₃₄	47 / 55	W. / S. / S.		vast dew. / bright. / summer's day.	Hyacinthus non scriptus	Regulus non cristatus medius. Bombylius major.			sings: a pretty plaintive note...
21.	Wednes. 8 / 12 / 4 / 8	29¼ ¹⁰⁄₁₀	52 / 54	W. / SW.		grey, mild, brisk wind. / sprinkling. / sun.		Alauda minima. locusta, voce stridet. Tryglus campestris		The titlark, a sweet song... not only sings flying in its descent, & on trees, but also on ground, as it... about in pastures.
22.	Thurs. 8 / 12 / 4 / 8	30 / 30	48½ / 57	NW. / NW.		great dew. / bright. / sweet summer weather.		Cuculus.		Grasshopper-lark chirps. Bat. Apricots begin to set. Cut first cucumber.
23.	Friday. 8 / 12 / 4 / 8	31 / 31	58½	NW.		wh. frost. / sweet summer weather. / dark & mild.	Prunus cerasus. Pyrus communis.				Four ring-ouzels appear the common: they feed on berries: are wild & shy: probably been shot at. A very late in their passage. Spring-corn comes up well.
24.	Saturday. 8 / 12 / 4 / 8	30	52 / 56	NW. / NE.		bright. / summer's day. / dark clouds.			Musca meridiana.		...much. Sowed carrots: planted po... Mild & still.

April 21. Field-crickets have opened their holes: they are full-grown, but have only the rudiments of wing... & are probably in their larva state; yet they certainly eat, as appears by their dung. It seems... ...ly, that they die every winter, leaving eggs behind them. About Septem.ʳ all the mouths of their ho... are obliterated. They do not cry 'till about the middle of May. Their noise is shrill & loud... This is by no means a common insect. They probably cast an other coat before their wings are perfect, & ... pond.

A page from Gilbert White's notes from April 18–22, 1773. White recorded his field notes in these standard-format journals that were published by Daines Barrington in 1767 as *The Naturalist's Journal* (the initial edition was anonymous).

Instructions for documentation were also given in letters. The third U.S. president Thomas Jefferson wrote to Captain Meriwether Lewis on June 20, 1803, with explicit instructions to notice all manner of plants, animals, and minerals on his westward journey, and suggested that "Your observations are to be taken with great pains & accuracy, to be entered distinctly & intelligibly for others as well as yourself" and "that one of these copies be on the paper of the birch, as less liable to injury from damp than common paper."[13] Even those field naturalists who stayed closer to home, such as Henry David Thoreau, kept careful field notes. In the 1850s, Thoreau received a circular from Louis Agassiz that described information he should record in his field notes on fish:

A notice of the physical character of the localities where specimens have
been collected would be a valuable addition to the collection itself. Respecting the land it should mention: the height above the level of the sea,
if known, the nature of the soil, whether dry, moist or swampy, muddy,
sandy or rocky, &c. Respecting the waters: the mean and extreme tem-
peratures, if ascertained, whether clear or muddy, and of what color,
deep or shallow, stagnant or current; of rivers especially, the rapidity of
the current, and also whether subject to great rise or fall.[14]

Since Thoreau's time, other systems for documenting natural history observations have been published for birds, insects, and general
natural history.[15] There is also a wealth of recent books on how to keep
"nature journals," which generally include sketches and basic observations.[16] Some field guides even offer simple instructions on how to keep
field notes. Even considering the materials available, a serious naturalist-
scientist is still left to ponder how field notes can be recorded efficiently
and effectively. The answer, clearly, is specific to the nature of the author
and the need addressed, which is why this book offers twelve different
opinions on these topics.

These myriad approaches to field recording balance certain common variables. Much can be learned from those who have worked to
keep careful records, and even Darwin's notebooks reveal this tension
between fact and theory, data and narrative. In many of Darwin's early
notebooks, such as his zoology notes aboard the *Beagle*, his accounts are
largely descriptive. He fills the pages with many observations and facts,
and his questions on evolution emerge between the cracks (as his con-
templation of the evolution of marine lizards quoted earlier illustrates).
Later notebooks, such as his infamous Red Notebook, move from observations to theory.[17] This notebook was started at the end of his *Beagle*
journey. In it he moves from documenting his field observations to considering the underlying principles on evolution that were fleshed out in
his subsequent notebooks.

Notebooks of modern field scientists still balance this composition-
al tension, and the continuum of information contained in field notes,
balanced differently depending on goals and discipline, can be divided
into several loose categories: diary, journal, data, and catalogs. Diary entries record information on mundane daily occurrences, such as meals,

of small fish which now begin to run and are
taken in great quantities in the Columbia R.
about 40 miles above us by means of skiming
or scooping nets. on this page I have drawn
the likeness of them as large as life; it
as perfect as I can make it with my
pen and will serve to give a
general idea of the fish. the
rays of the fins are boney but
not sharp tho' somewhat pointed.
the small fin on the back
next to the tail has no
rays of bone being a
= branous pellicle.

to the gills have
each. those of the
eight each; those
are 20 and 2
that of the back
the fins are of
is of a bleuish
the the lower
is of a silve=
part. the
behind the
second of
the puple
a silver
and
like

see other R.
h. 51

thin mem-
the fins next
eleven rays
abdomen have
of the pinnaani
half formed in front.
has eleven rays. all
a white colour. the back
duskey colour and that of
part of the sides and belly
ong white. no spots on any
first bone of the gills next
eye is of a bleuis cast, and the
a light gaald colour nearly white.
of the eye is black and the iris of
white. the under jaw exceeds the uper;
the mouth opens to great extent, folding,
that of the herring. it has no teeth.
the abdomen is obtuse and smooth; in this
differing from the herring, shad, anchovy;
&c of the Malacapterygious Order & Class
Clupea

expenses, and meetings with others; journal accounts include weather conditions, daily movements and geographic locations, and basic observations of plants and animals. Data entries encompass substantial behavioral observations, factual records, and experimental results; and catalogs record things collected and observed. Although their boundaries are porous, such categories are useful when examining how field notes vary. In some disciplines such as systematic collecting, cataloging information on species and collections may be primary and other ecological information deemphasized. In more empirical undertakings such as ecological studies, the balance of content may shift to elements of experimental design and data, with these composing the majority of the notes. In paleontological notes, records of particular facts and locations of objects can be essential. Certainly, the authors in this volume consider the ways that field notes are cohesive documents composed of facts, theory, data, and narrative.

Across disciplines, a related balance is struck in how information in notes is organized. Some pursuits accommodate a free-form approach in which ideas range widely, whereas others require consistency and standardization. Some types of information find their home in bound journals while others rely on uniform field cards and data sheets. Recording diverse types of data is possible in our technological era, but determining the best method of doing so still requires thought and a long view of one's goals. One persistent tension in organizing field observations is that some of these categories are inherently chronological and some are not. Diary and journal information fits a daily protocol, but data and experiments may be collected intermittently over large periods of time and may not be relevant to the diary entries of those same days. Darwin kept small field notebooks and a diary along with dedicated zoological and geological notebooks. The authors in this book provide a range of different methods—from systems with a free-flowing journal to those with dedicated diaries, journals, and catalogs. Many keep a small pocket notebook in which they jot small notes during the day that are then transcribed more fully, as Darwin did, into formal journals.

(opposite) **Meriwether Lewis's journal notes of the Eulachon fish (*Thaleichthys pacificus*), made on February 24, 1806, while Lewis was near Fort Clatsop, Oregon.** Used by permission of the American Philosophical Society.

Modern field scientists may pursue a variety of organizational solutions for integrating the information collected in the field, from the paper and pencil method that arose from the followers of Joseph Grinnell to relational databases.

As field workers address these issues of content and organization, they must also consider the ultimate value of these notes in relation to their objectives. Since human memory is transitory and things that are not written down may slip away quickly, field documentation is critical. However, there is clearly an opportunity cost to taking field notes. Every minute spent taking notes is a minute that could be invested in something else. Experiments, specimen preparation, and sleep often take priority over making notes, and since one cannot possibly record everything, an appropriate level of investment in notes can be essential for the success of fieldwork. Certainly, some eminent field scientists have been successful without keeping integrated notes at all. Regardless, deciding how much energy to put into field notes requires determining what information is worth recording.

The value of taking field notes lies both in the actual information that is recorded as well as in what is gained in the process of recording itself. Darwin's collection lists that accompany his zoological notes still have scientific value in that they describe exactly what he collected and where. His records of observations provided information he later required to write his *Voyage*. Field notes provide written records both for the scientist and for future generations. Careful records on experimental design and theory can be scoured to reveal possible mistakes or missteps, or for protocols that allowed for important discoveries. Location data provide specific information on how to find organisms centuries in the future. It is impossible to predict the future relevance of any one page of notes. Yet it is clear that meticulous and organized records form the foundations of field science, and, like laboratory notebooks for our indoor relatives, are the most basic tool for studying the science of nature. Although the content of field notes has incredible value, the act of recording field notes has benefits that are less apparent and often underestimated. Darwin's field notes, for instance, proved indispensable for the information they contained, but did they also force him to reconsider previously formed ideas?

Charles L. Hogue   COLLECTING NOTES   No. CLH 1649

Locality 2 mi. S. Little Rock Dam (M) San Gabriel Mts.   Coords. 118°1'-34°27'
District CALIFORNIA   Sub. Los Angeles County   Country U.S.A.
Date 15 May 1965   Time 1 P.M.   Elevation 3800'
Collected by C.L. Hogue   Method net

SITE

TERRESTRIAL Visiting flowers of Salix sp. growing by small, sluggish stream

Weather almost clear, hot day (third in a series of three following mild storm)
Temperature 85 deg. F.   Humidity 16 rel. %   Barometric pressure 29.6 rise ✓ fall
Clouds sparse, high   Wind intmt. breezes force 0-10 mph direction S - SSE
Terrain gravelly, boulder strewn wash slope level % direction

AQUATIC Small, ground pool in a shallow depression beneath edge of a large Baccharis shrub (P)

Size oval, 6' x 2'   Flow none - stagnant
Salinity none (by taste)   Other solutes not determined
Temperature 73 deg. F.   Color clear   Surface light bacterial scum
Bottom algae covered granite rocks   Shade partial
Vegetation abundant & thick masses of Spirogyra; sparse grass near edge

ANIMAL HOST wood rat
Species Neotoma fuscipes macrotis det. C.L. Hogue
Age adult   Size 388-191-39-32   Sex ♂
Situs base of tail   Preserved: yes ✓ no   Museum LACM no. 28621

OTHER

GENERAL ENVIRONMENT

Artificial
Natural Shadscale Scrub (Munz & Keck)

COLLECTIONS

No.	Identification	Remarks
A	Autographa californica	♀ - confined, laid 60 eggs (over)
B	blue megachilid bee	exhibited peculiar feeding behavior *
C	large tachinid	
D	Andrena sp.	
E	Culiseta incidens	2 blooded ♀♀; both confined, #1 laid 30 eggs, #2 " 40 "

☑ *See supplemental sheets for additional notes.   ✓ (P) Photographs   ☑ (M) Maps

FIG. 1.—Field-note form, front. Hypothetical examples are inserted for all categories under "Site." Under actual conditions, only 1 category would be completed for each collection.

An example of a standardized field-note form for insects presented by Charles Hogue in his paper, "A field-note form for general insect collecting" (Hogue, 1966). Used by permission.

Taking time to write out an idea or observation forces us to pause and consider. Recording the daily unfolding of experiments—their success or failure—encourages an honest assessment of how each day's work fits within the underlying goals and theory of the project. It takes time to create a narrative of experiments, events, and observations, but it eventually pays dividends because it forces thorough examination, which is a common characteristic of science across disciplines. In Darwin's description of the marine iguana, for example, we can imagine him on board the *Beagle*, penning his zoological notes and pondering the origin of their "apparent stupidity."

In contrast to these seemingly timeless dilemmas about the value of records, other challenges have presented themselves with the rise of technological solutions for collecting information in the field. The use of many kinds of digital media have made keeping field notes both easier and more complex. Computer sensors, handheld devices, and digital cameras and microphones can all capture huge quantities of information in seconds, but these volumes of unstructured information are not cohesive field notes, though they may provide a false sense of completeness. Such data are not naturally integrated, and are often scattered among multiple devices, each requiring specific technology to access. The raw information lacks both a narrative and a record of how and where information was recorded. Providing this record is the role of field notes. When deciding how to record work in the field, consider this: Are there documents that explain what, how, and where things happened that are accessible to an independent reader?

Certainly technology plays a role in keeping field notes. Many field workers find a way to transfer their notes to an editable format in a digital medium. Emerging technological applications include digital pens that record a duplicate virtual copy and digital journaling software.[18] Relational databases allow data and virtual notes to be electronically linked for quick and powerful access and searches. Whether our pens are digital or ballpoint, however, the goals of keeping field notes remain unchanged.

Authors in this book consider different technological approaches to field notes, and these varied perspectives raise questions about what may be gained or lost with the implementation of digital notes. What are the differences between an entry recorded in a word processor or a digital camera and one recorded as written text and sketches? What might a young naturalist who records notes in a blog or digital slideshow learn from seeing how an earlier generation took notes? Are there elements of science that are more thoroughly documented when recorded by hand? Regardless of how they are created, the tradition of creating field notes remains of critical value to scientists and natural historians.

The objectives of most field workers contain elements of both science and natural history, but in different measures depending on each field worker's research goals. The evolutionary biologist Naomi Pierce once recounted how Bert Hölldobler, the eminent sociobiologist, had encouraged her to not simply record observations but to focus on quantifying them. If we are coming to the field with the aim of investigating interactions empirically, it is incredibly valuable to record information systematically in data sheets and notes so that it can be subject to rigorous comparisons later on. In contrast, if we are approaching the field to make general observations or study a new fauna or flora, we may instead benefit from the open forum that a blank journal page provides. I don't intend to referee between various approaches or to determine the nature of "real" field notes. Instead, it is my hope that the perspectives offered in this book will provide choices and encourage consideration of other systems. Aspiring field scientists might do well to consider how recording diary information, in addition to keeping data sheets, might serve them when undertaking the rigorous accounting often required of scientific grants, or how a journal might provide a wide-angle forum for more general reflection on experiments and organisms, or even how it might have personal value later on when reflecting on their adventures in the field. Similarly, those who more closely follow the tradition of the historical naturalists might benefit from reflecting on how an increased emphasis on quantified observations might make their work more powerful.

Ultimately, this book allows anyone a chance to peer over the shoulders of outstanding field scientists and naturalists and into the pages of their journals. These specific examples are methods that can be adopted wholesale or tweaked to fit a multitude of agendas, and can also be used as starting points for anyone interested in the natural world. These authors raise both unique and universal issues that emerge across disciplines, though the individual quirks, eccentricities, and real-life adventures they record are part of what fit these documents into the broader topic of doing science. Taken together, they emerge at the intersection of person and place to reveal how naturalists think and work in the field.

The tradition of field notes that grew into its own genre over the past three centuries is still relevant to anyone who studies nature. Although the diversification of field pursuits and the complexity of their studies have expanded the scope and methods for field documentation, the basic role and importance of field notes are unchanged. The examples, thoughts, and instruction provided in this book are only a first step in maintaining the valuable tradition of field notes, and are meant to encourage more rigorous and long-lasting documentation of our natural world.

1

The Pleasure of Observing

GEORGE B. SCHALLER

The Scientist does not study nature because it is useful to do so. He studies it because he takes pleasure in it; and he takes pleasure in it because it is beautiful.

—*Jules Henri Poincaré*

THE LION PRIDE, consisting of three maned males, seven females, four large cubs, and six small cubs, finally stirs itself after hours of indolence. It is shortly after midnight, and a moon suffuses the Serengeti plains with silver so bright that the lions cast shadows. I use no artificial light to observe the animals. It might disturb their potential prey. Nothing violates the vast silence except the distant whoop of a spotted hyena. Brittle grass crackles beneath the lions' heavy tread. One lioness, a little apart from the others, begins to dig at a warthog burrow, sweeping sandy soil backward first with one paw and then the other, claws unsheathed. Two lionesses join her and excavate intermittently; the rest of the pride lounges nearby. After an hour the lionesses have exposed some two and a half meters of tunnel, which runs level and 0.6 m below the surface. Suddenly, a lioness ducks her head into the tunnel, grabs something, and strains, her powerful shoulder muscles bulging. Two males walk up and watch expectantly. Another lioness continues to dig, engulfing the site in a dust cloud. After maintaining her grip for eight minutes, the lioness suddenly pulls hard and drags a squealing male warthog by the nape from the burrow. The whole pride rushes and covers the warthog with squirming and snarling bodies. The air is heavy with the odor of blood and stomach contents. By 2:00 AM only two males still squabble over the warthog head. The rest lick themselves and each other or gnaw on remains, teeth grating on bone. Toward morning the pride roars on

8' dug up

3' not dug

entrance
20"

Pig hole

Pig burrow was tunnel about
11 feet long with bottom about 2'
below ground and a roof 6-8" thick.
It ran almost level and straight
except for bend at the end. The lions
removed the whole roof for about
8' of the tunnel

see
Tape
III

0135 ♀ still digging   34

0137 Several times ♀ had put head
into hole and jerked it back
suddenly. Now she reaches in and
grabs something. Her head is out
of sight and she is straining.

No 2 ♂ arrived within past half hour.
Now he and No 3 ♂ come up and
stand by ♀. The other ♀ who had
been digging paws at the hole
entrance, creating a dense cloud
of dust.

After 8 min of holding the head in
the hole, the ♀ starts to pull.
A ♂ warthog squeals, she seems to
have it by the nape. As she pulls
it out the 2 ♂ grab it too. Rest
of pride rushes up and the pig
is covered so completely that it
is invisible under the mass of lions.
3 small cubs eat too.

0200 Only the head has meat left.
The animals are scattered with
bones gnawing. The 2 ♂ squabble
over the head. While ♂3 lies on
his side, ♂2 is draped over his
neck, each retaining his hold.
They keep this position for 5
min., then ♂3 gets head when other
gives up. He takes head 100' and
eats.

Notes from October 29, 1966, record how a Serengeti lion pride digs a warthog out of its burrow.

seven occasions, and I record the thunder of their communal voices on tape.

At 5:10 AM, I drive back to our bungalow, where my wife, Kay, and our two small sons live. I had learned during the past twenty-four hours that the pride walked four kilometers and made three desultory stalks and chases, once after a reedbuck and twice after gazelle. I had also noted who interacted with whom, and, in general, obtained facts about the daily routine of a pride. During our three and a half years in the Serengeti, I collected thousands of discrete pieces of information on pride movements; the age, sex, and physical condition of each prey species killed; and social responses of individuals within and between prides. I collected these data to answer a question: What effect does lion predation have on prey populations? Such research consists of a repetitive recording of the same facts until one has a large enough sample to deduce patterns, derive

Details of two separate lion hunts, both unsuccessful and not very determined. The upper drawing shows two lionesses stalking toward a Grant's gazelle, and the lower one shows three lionesses approaching a reedbuck.

conclusions, and even make predictions. Lions are most active at night when their daytime lethargy vanishes. Therefore, collecting these data involved spending long days and even longer nights watching the animals.

On a typical day I would observe the lions of the Serengeti while sitting, usually alone, in a Land Rover, an accepted and innocuous presence near the animals. One must be close enough to observe, but not so close that it affects the routine of the animals. Binoculars, spotting scope, notebook, and pen or pencil were my basic tools. I scribbled into a pocket notebook as an incident occurred or immediately afterward, often jotting down just a few key words without looking at the paper, especially at night. It was essential to write notes quickly because memory is notoriously imprecise. I made small maps of routes and approximate distances taken by animals, such as when lionesses fanned out to stalk prey. Simple sketches of postures helped me to visualize the animal later. Check sheets were valuable for tabulating activity readings from a radio transmitter or other such simple records, but I generally do not like to abbreviate behavioral notes. An important detail may be ignored or considered irrelevant and discarded because it lacks a discrete category on the list. It is often an anecdotal event that offers special insight.

When I returned to camp, the rough field notes were soon transcribed legibly and in greater detail into a permanent notebook, a task that could take more than an hour. I also added comments and ideas when I copied the notes. I used hardcover, sixty-page notebooks, either 12 × 19 or 20 × 25.5 cm in size, as I still do today. The clarity of these notes was essential because they were later used as the basis for scientific papers, where the individual observations were summarized, quantified, and interpreted. Such transcription has a further advantage in that it provides a duplicate set of notes in the event that one is lost; I keep the two sets in different places. These final notebooks on lions are permanent, indexed, bound records available to anyone. They now are part of my collection of about three hundred of these notebooks based on wildlife research in twenty-three countries. Some projects consisted of surveys of a month or two, such as in the rain forest of Vietnam in search of the rare Javan rhinoceros, or in Sarawak on an orangutan census. These might require only one notebook. Long projects, as on lions or giant pandas, may fill a shelf in my study.

I also keep a personal journal, separate from my scientific one. It is a daily record of impressions, ideas, concerns, and complaints. I describe the doings of my family and of my activities, make comments about other people, and relate matters beyond what is pertinent in a scientific notebook. It is also a place for displaying emotions—the joy of discovery, the beauty of a giraffe at sunset, the deep pleasure of observing the rich life of another being. Such notes contribute to later popular writing, creating empathy between reader and animal.

It may seem anachronistic that I write notes in longhand in this technological age. I do not use a tape recorder to take notes, and the laptop had not yet been invented when I began field work. A machine loses data too easily; it has to be maintained under rugged conditions; it is yet another item to carry; it tempts thieves. However, I do use technology as it becomes available, such as GPS and radio-telemetry, if it enables me to obtain data more easily. I consider all machines mere tools. My main focus is providing a species with a written history, a biography, and for that one must join the animals. Looking at a computer screen, flying over terrain, and plotting locations on a map as relayed by satellite may all add to the mountain of available information. But none provides the pleasure or the knowledge and experience of being with the animals—of getting your boots dirty, as they say. Konrad Lorenz years ago bemoaned "the fashionable fallacy of dispensing with description" of an animal's behavior. Luckily, I have had the pleasure of spending much of my career in the field in direct contact with animals—watching, listening, and writing.

These close encounters have allowed me to develop an understanding of snow leopards, jaguars, and other secretive creatures. At times, I have tranquilized, handled, tagged, and radio-collared these animals. But I am aware that such intrusion into their complex lives could harm them emotionally and physiologically, and even kill them if they are carelessly trapped or drugged. Given my respect for animals as individuals and my responsibility for their well-being, I capture animals only with great reluctance. More often, I have been amazed at how much I could learn by instead being a patient and persistent observer.

In every project there is pressure to amass quantitative data, thereby justifying financial support and enhancing one's scientific credentials. To collect this information in rigorous ways, you want to be able to identify animals as individuals in any behavioral study, if possible. Scars, pelage patterns, and other minor variables, singly or in combination, help in such identification. I could easily recognize a tiger by the distinctive stripes on its face. If a species is social, then comparisons among individuals may be even easier. Mountain gorillas, for example, have such a distinctive nose shape and wrinkle pattern that I knew over a hundred animals by name. Such recognition raises a project to a new level. Individuals not only become acquaintances whose lives one follows avidly, but also provide details of their society that would otherwise remain obscure. If one has a personal knowledge of the individual animals being studied, observations in field notes cease to be impersonal, and an observer's empathy can lead beyond dry facts to better intuition and insight.

After all, every animal is an individual with a personal history, its

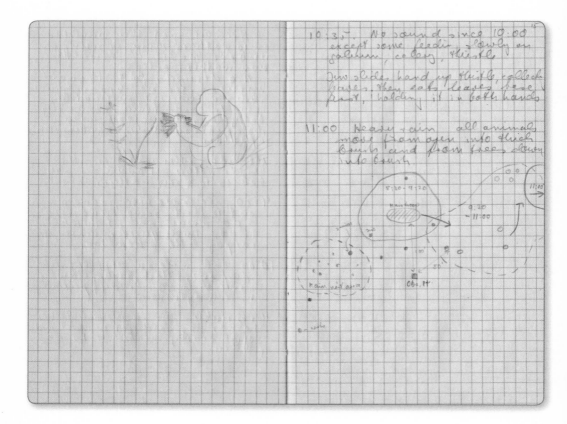

Tracking mountain gorillas on January 12, 1960, in what was then Parc Albert, Belgian Congo (now Virunga National Park in the Democratic Republic of Congo).

behavior influenced by friends and enemies, relatives and neighbors. Since we cannot interview the subject, we can only infer the past from the present, realizing that conclusions may be skewed. Ideally, a study should persist for at least the life span of an animal—perhaps fifteen to twenty years for lions and thirty to forty years for gorillas. My research on each of these two species was a brief two to three years, but others have continued the projects since the 1960s. In pioneering efforts such as mine, one grasps at all information rather indiscriminately, taking notes on anything easily observable, such as food habits and the daily routine. Previous studies on other animals provide guidelines, but every species is unique and demands fresh perspectives, a knowledge of new individuals and of the meaning of unique behaviors. Concepts in science evolve and new questions continually obtrude, and so more extensive studies emerge from pioneering ones. Compare my book *The Mountain*

Gorilla (1963) with *Gorilla Society* (2007) by Alexander Harcourt and Kelly Stewart, and note the vast difference in approach and detail in the latter. In my work, I have been continually drawn into studies of enigmatic species about which little is known. Part of this may simply be who I am, but some of my interest in uncovering and documenting these species was inspired by the work of the early ethologists.

I was influenced in my outlook on fieldwork by three great naturalists. Those with a passion for animals often translate their observations into superb popular accounts to convey their enthusiasm and stimulate that of others, and one can easily get a sense of the ideas that shaped my approach by reading *King Solomon's Ring* (1952) by Konrad Lorenz and *Curious Naturalist* (1958) by Niko Tinbergen. These works make clear the emotional depth with which these two founders of ethology viewed the natural world, and their passion quickly rubbed off on me. In 1956, I was offered the opportunity to be a field assistant on an expedition led by Olaus Murie to Alaska's Brooks Range, an area that as a result of our work became the Arctic National Wildlife Refuge. It was a wonderful summer spent compiling a bird list, collecting plants, and making an inventory of the natural riches in general. Although in his late sixties, Olaus approached each day with curiosity and a sense of wonder, and he conveyed to us what he called the "precious intangible values" of this wilderness. As I look back at my notes from years of fieldwork, I realize how these experiences made a deep impression on me. Learning how to observe and document species and also to appreciate these "intangible values" charted a course for me in science and conservation that I still follow.

My early detailed studies were primarily in national parks, and they focused on the natural history of species. Conservation problems are social, economic, and political, and they were then secondary to developing an understanding of species' natural history. This has changed. Now I collect data mainly to help protect and manage the species. The Tibetan antelope, or chiru, provides a good example. This elegant animal migrates across the Tibetan Plateau of China mostly at 4,300 meters in elevation and above. It roams over an area of half a million square kilometers, an area larger than Texas. Chiru have been hunted heavily for their fine wool, called *shahtoosh*, and this has left their populations in jeopardy. Obtaining the basic information needed to conserve this

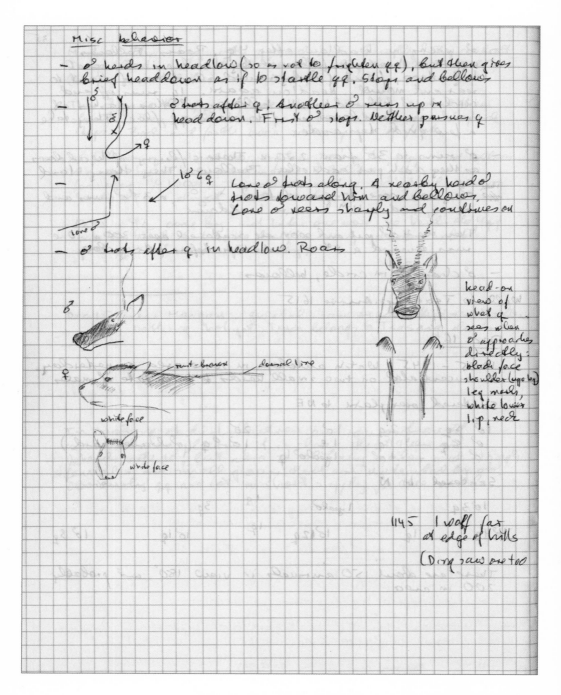

Behavioral notes on the chiru (Tibetan antelope), a species that ranges over half a million square miles of the Tibetan Plateau. December 16, 1991.

Chiru birth   July 3

snowing to 1130 but snow melts off S-facing slope by early afternoon. 1600 snow flurry
1700 hailstorm, 1800-1900 hail → snow storm

♀ is on open lower slope, about 30 m from bottom, perhaps 15° slope but on a small more level
spot. I am nearly 1 km away with scope. Heat waves and distance make me miss some
detail.

1435 I see ♀ lie on side, hindlegs shielded, but head raised
1441 ♀ stands up, then lies
1445-46 ♀ stands and lies 3x
1447 ♀ stands, newborn is hanging out, head nearly touching the ground. ♀
      turns 2x 180°, fetus swinging and lies down. She stands immediately
      and I can see newborn struggling on ground. ♀ seems to lick it and her tail
      wags fast
There are at least 7 other chiru nearby, the nearest at 10 m, foraging but
none respond
      young rears on forelegs. ♀ seems to nuzzle it
   1451 ♀ lies. Young struggles by her belly
   1455 ♀ seems to lick young. young continually rears head up
   1502 young gets up on its 4 legs   (15 minutes after birth)
         ♀ stands up then lies again. Young gets on 4 legs, stumbles backward
         and falls on rump
   1505 ♀ stands up. Young does too but falls on rump
   1506 ♀ lies. Young stands up, lies down. ♀ stands up, lies down
   1510 ♀ stands and seems to lick young
   1511 young stands and seemingly tries to suckle - head around belly
   1511 Young takes several steps. ♀ lies   ♀ lies down, young walks by ♀
   1515 Young takes several fast steps
   1520 ♀ stands up and feeds briefly then lies ; young lies
   1521 Both ♀ and young stand up. Young nuzzles around her belly
         ♀ feeds again. Young takes several steps with her and collapses
         ♀ lies again. Young walks 2/3 around her and lies close to her
   1525 ♀ stands and forages. Seems to lick young.
         Young apparently suckles 1.5 min, standing at right angles to
         her, its muzzle in her groin. She stands still (38 min after birth)
   1530 Young lies as ♀ feeds, ♀ lies. Young walks to her head and lies
   1535 ♀ feeds again, then lies. Young circles ♀ and goes to her head of
         ♀ stands and seemingly lick young. young lies, ♀ feeds
So far all action at birth site about 3 m diam.
         ♀ moves about 3 m from birth site and lies. Young walks to her head

                                        cont p.27

Detailed field notes on the birth of a chiru. July 3, 2005.

widely ranging animal requires expeditions into remote places, many of them uninhabited.

My local coworkers and I take notes on the location, size, and composition of herds; birth and death rates and causes of death; migration routes; seasonal food habits and nutritional contents of plants; competition between wildlife and livestock; and other various topics. We collect

Diameter of shoots eaten by Zhen   May 31 AM

1.39	1.14	1.11	1.30	1.21
1.22	1.65	1.29	.98	1.38
1.70	1.33	1.25	1.58	
1.28	1.35	1.66	1.07	Total 102
1.05	1.78	1.40	1.17	13965
1.34	1.46	1.71	1.21	Mean 1.37
1.35	1.51	1.49	1.00	
1.39	1.37	1.15	1.35	
1.03	1.18	1.10	1.11	
1.30	.91	1.06	1.43	
1.32	.99	1.42	1.50	
1.28	1.11	1.12	1.29	
1.29	1.58	1.24	1.30	
1.40	1.49	.92	1.56	
1.13	1.15	1.75	1.21	
1.48	1.48	1.54	1.28	
1.32	1.90	1.55	1.36	
1.04	1.40	1.74	1.22	
.96	1.54	1.76	1.44	
1.67	1.63	1.60	1.50	
1.65	1.18	1.51	1.45	
1.32	1.23	1.40	1.33	
1.73	1.40	1.77	1.55	
1.31	1.75	1.55	1.10	
1.58	1.81	1.35	1.38	

Dropping shoot		
6	1030 gm
5	770
5	910
5	1020
4	650
6	880
4	660
35	5920 g

169.14 g per dropping

Indirect observations on the giant panda in the Wolong Nature Reserve, Sichuan Province, China. My notes record the diameter of bamboo shoots eaten by a panda and the wet weight of droppings.

wolf scats to identify prey items, whether chiru, livestock, or marmot. It is in some ways a superficial project, huge in scale, unlike my intensive observations on lions and gorillas, but the results have led to conservation initiatives. I have also become connected to these austere uplands where horizon gives way to horizon, where sometimes a chiru herd flows across the landscape in a throbbing stream of life. And there are moments of intimate pleasure: one December day the chiru danced near me in their mating ritual, offering detailed notes on behavior for comparison with related species; on a snowy day in early July, a female gave birth on a barren hillside.

In 1980, I was asked to assist with a study of another little-known species, the giant panda, and spent four years in the fog-shrouded mountain forest of China's Sichuan Province. The panda is rare and an icon of conservation, and facts were needed to protect it and its habitat. The first task of this Chinese collaborative project was to describe the panda's way of life, which was not an easy goal because the animals live in dense and dank expanses of bamboo. With direct observations difficult, we concentrated in part on the artifacts of a panda's passing—feeding sites, droppings, scent posts. We noted the species of bamboo eaten and the food type selected, whether stem, leaf, or shoot. The height and diameter of shoots were measured and recorded to detect preferences, and I used this information to make maps of their movements. We examined droppings and sorted the poorly digested bamboo into shoots, stems, and leaves to determine the proportion of each consumed. Bamboo was analyzed for amino acids, vitamins, crude protein, cellulose, and other constituents in a laboratory. We also followed pandas in snow to count and weigh droppings. For example, we tracked one male for five and a half days, and he deposited an average of ninety-seven droppings weighing 20.5 kg per day. Pandas obviously eat a lot of bamboo.

We radio-tracked several pandas to determine the size of their home ranges (3.9–6.4 sq km) and their cycle of daily activity. By monitoring signals from certain individuals every fifteen minutes for as long as twenty consecutive days, an exhausting task, we learned that pandas are active, mostly eating, for 14.2 hours per day, in daytime as well as night. Our activity readings totaled 28,450.

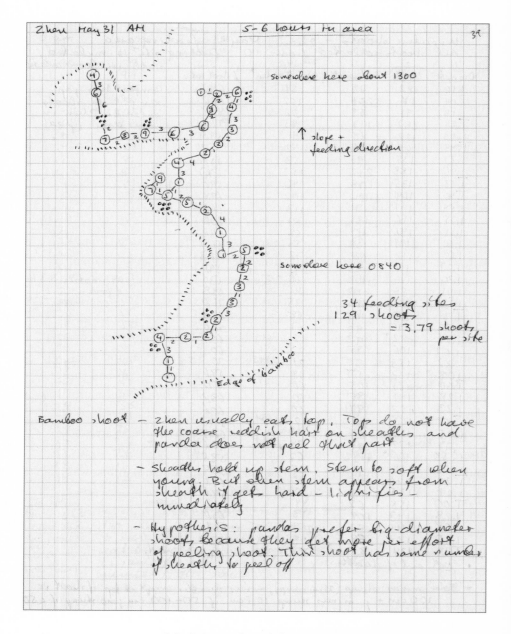

Zhen May 31 AM — 5-6 hours in area — 34

somewhere here about 1300

↑ slope + feeding direction

somewhere here 0840

34 feeding sites
129 shoots
= 3.79 shoots
per site

Edge of bamboo

Bamboo shoot — Zhen usually eats top. Tops do not have the coarse reddish hair on sheaths and panda does not peel that part

— Sheaths hold up stem. Stem is soft when young. But when stem appears from sheath it gets hard — lignifies — immediately

— Hypothesis: pandas prefer big-diameter shoots because they get more per effort of peeling shoot. Thin shoot has same number of sheaths to peel off

A detailed route of a panda foraging on bamboo shoots, showing the number of shoots eaten and droppings deposited (black spots) on May 31, 1982.

In spite of pertinent conclusions about a panda's energy budget and lifestyle, our work had limitations. An imaginary panda admonished us about this, as related in my book *The Last Panda* (1993):

> Honorable scientists: I want to compliment you on your efforts to study my kind. It takes dedication of the highest order to measure so many droppings. Day after day you follow my tracks with admirable persistence if not technique: I can hear and smell you from far away. Actually I'm not certain what you expect to gain from invading my privacy. You generate numbing statistics about the number of stems I eat in a day and the number of hours I sleep . . . it merely shows that you have discovered some easy facts about me; most aspects of my life cannot be written in the language of mathematics. How can you understand me? We may seem to share certain moods, but you cannot comprehend mine. After all, it's not your perception of reality that matters . . . Another point. You study my diet, you study how many times I scent mark and mate and how far I travel. Remember, you cannot divide me into independent fragments of existence. At best you might perceive an approximation of a panda, not the reality of one. I am, like any other being, infinite in complexity, indivisible, a harmonious whole . . . we shall always remain of two worlds. Humans can never know the truth about pandas. Therefore, enjoy the mystery—and help us endure.

Today I cannot look at animals like these pandas without feeling sympathy, concern, and a sense of guilt about their potential fate. Anyone who observes the destruction of wildlife and its habitat must become an advocate for conservation. We must all live by a land ethic, as Aldo Leopold urged in *A Sand County Almanac* (1949), in what is to me the most influential and beautifully written book on conservation. There is still so much work to be done to understand the natural history of all species and habitats, not just endangered ones. Conservation initiatives begin with recording observations. After all, field notes on droppings do provide useful insights.

2

Untangling the Bank

BERND HEINRICH

IT ALL STARTED when I was eight years old. I began running along the gravel roads near my childhood home, keeping my young eyes open for beetles and birds. By the time I was in high school in Maine, I had graduated from going barefoot most of the time to wearing a pair of canvas shoes with hard black rubber soles. This official running attire transformed my running into a more serious pastime when, as a junior, I made it onto the roster of our high school team. To establish rewards for myself, I purchased a pocket-sized spiral notebook that served as a trophy log where I wrote down the results of our cross-country races. The notes I kept to mark my accomplishments were simple. They documented the distance and time of my runs, along with who ran and how they placed. I started running each day, and my distances increased to four or five miles. On these runs my eyes and mind continued to be attracted to the plants and animals dancing across my field of vision, and on the reverse side of my running notebook I recorded trophy sightings on my other emerging love, that for natural history.

I had been collecting specimens since I was five, especially carabid beetles, which I pinned into a box rather than recording the memory of them on paper. I soon hunted birds for my father with my slingshot, and we sold them to the Museum of Comparative Zoology at Harvard and other institutions. My animal encounters acquainted me with the nomenclature and habits of local species. Eventually, I started taking baby birds from nests and raising them as pets (never caged), most memorably a crow, a wild pigeon, a hawk, and a jay. Several years later, at our

Hinckley, Maine

1957

Apr. 21 – Barred Owl eggs ready to hatch

Apr. 28 – Red shold. Hawk eggs slightly hatched no leaves on trees

May 11 – White Breasted Nuthatch eggs strongly hatched. Leaves on trees

May 12 – Partridge eggs slightly hatched. Trees fully leafed out choke-cherry blooming. Sparrow Hawk eggs just laid.

May 18 – Brood winged Hawk, eggs moderately hatched. Apples blooming.

May 20 – Baby Robins just hatching. White Throated Sparrow building. Bobolinks back.

May 22 – Cliff, Tree and Barn Swallows building. Baby crows starting to get feathers. Flicker already made hole; sapsucker still making hole. Phoebe just laid eggs.

May 26 – Ovenbird building ~~and chipping sparrow~~.

These pages are from my first field notes kept in a small notebook while I was at the Good Will Home and School in Hinckley, Maine. My natural history notes were kept from front to back. I was seventeen years old, and my English was still very rusty. By "hatched" I meant *incubated*.

(continued on next page)

farm in Maine, I built birdhouses and hung them in the sugar maple trees next to our house so I could watch and listen to a variety of occupants that took up residence there.

As my fascination with bird life grew, I started keeping entries in a small spiral notebook of observations on the birds I had seen, perhaps so I could better predict when I would find them again. These notebooks expanded from ornithological phenology to flowering dates of plants, and they helped me, more than a calendar, to predict when to expect these occurrences again. During my somewhat overwhelming college years, I stopped such informal note-taking entirely. But as the years passed it resumed, and my notebook entries eventually contained increasingly detailed information.

About ten years after college, I again started running, this time with particular goals in mind. I recorded daily running distances

Cross-Country
 meets — 1957

① Good Will
   Waterville J. V.
        Came in 3rd
   for team, 5th for race
   out of 20.

② Waterville
   Good Will
   Lawrence (Fairfield)
   Winslow
   Guilford
        Came in 3rd
   for team, 12th for
   race out of 60.

③ GW -16 ; Rockport -41
   Heinrich   GW
   Salisbury GW
   Pottle   GW
   Hillard   GW
   Merrill   R.
   Broke course record of
   12:36 in 12:05.

④ GW-19 ; F.S.T.C. - 44
   Heinrich   GW
   B. Budzko F
   E. Salisbury GW
   D. Hillard GW
   B. Pottle   GW
   Broke course record of
   14:3¾4 in 14:30.

in my notebooks as rewards, symbolic trophies of what I had done to stay motivated and on track. Having specific aims, I then started to add almost everything else I could think of that might help me monitor and control my progress. This included how I felt, what I had eaten, the pace, my endurance and speed of recovery, my mental attitude, and anything else that might make a difference in how well I ran. I had noticed that on some days I felt that I could run like the wind, and on others that I could barely crawl. I kept data in my notebooks in the hope that they would help me understand the biology that might account for these swings in performance. This documentation made me a better runner by increasing my knowledge of how my body works under different conditions, and I have now enjoyed running for over fifty years, and have at times competed in races.

(State of incubation was determined by external appearance of the egg and/or quick immersion in a can of water—fresh eggs sink, incubated ones float.) My notes on my cross-country races appear on the reverse pages, from back to front. Here is the first page of notes on cross-country "meats" (sic) from 1957, and also results from several races in 1958.

May 1, 1962. I had taken a year off from my undergraduate studies at the University of Maine in Orono to join my parents on a bird-collecting expedition to Tanzania. I was a hunter/taxidermist and seldom had time to write, but I made occasional journal entries because I wanted to keep memories of places I thought I would never see again.

In this entry I described my euphoria of being out of the jungle and free in the acacia steppe near the village of Same, where I felt very much at home.

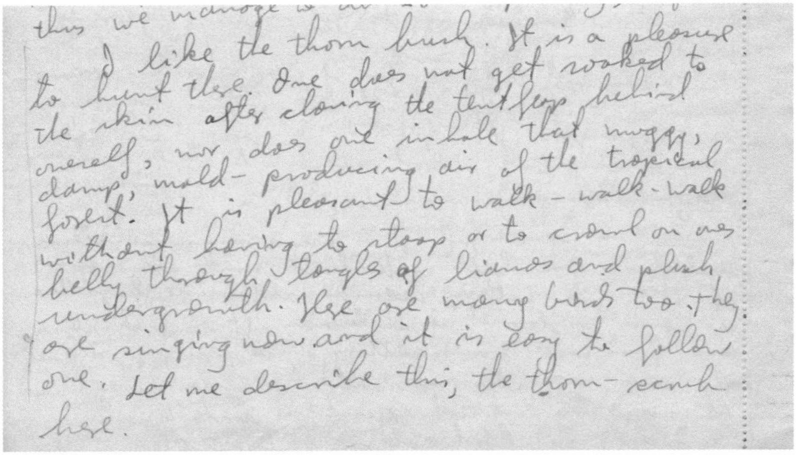

[handwritten journal entry:] ...this we manage to ... ... I like the thorn bush. It is a pleasure to hunt there. One does not get soaked to the skin after closing the tent flap behind oneself, nor does one inhale that muggy, damp, mold-producing air of the tropical forest. It is pleasant to walk – walk – walk without having to stoop or to crawl on one's belly through tangles of lianas and plush undergrowth. There are many birds too, they are singing now and it is easy to follow one. Let me describe this, the thorn-scrub here.

[handwritten field notes:]

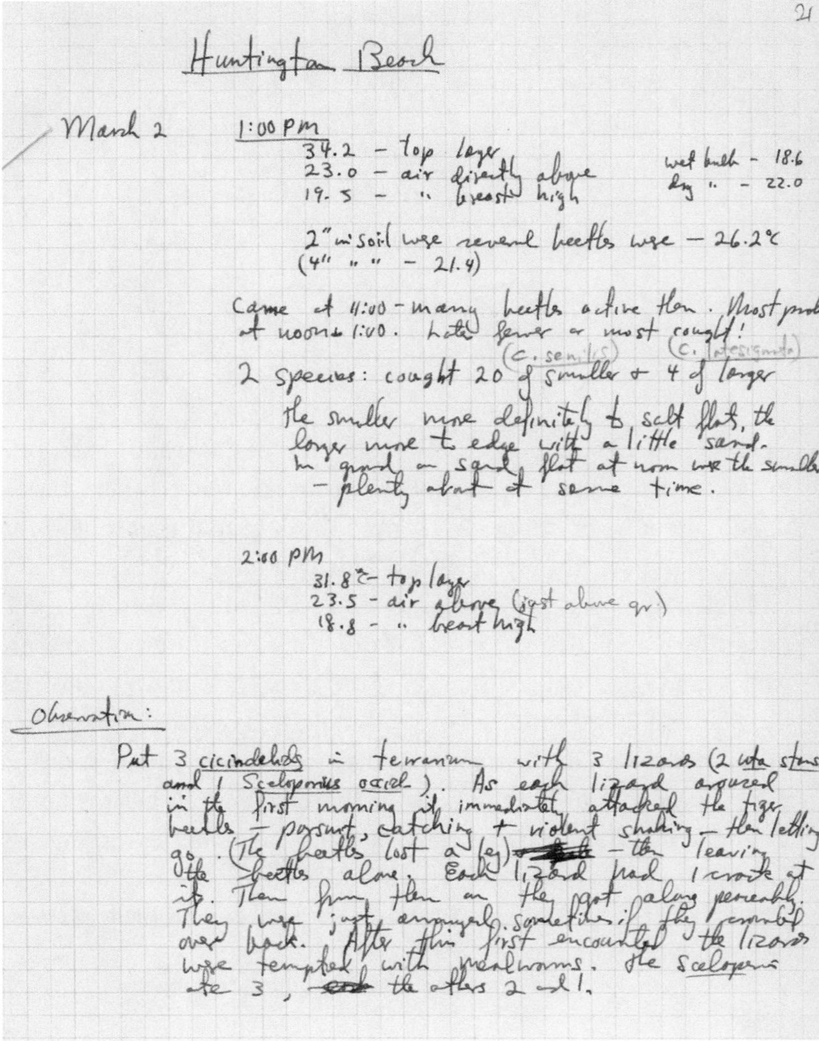

21

Huntington Beach

March 2   1:00 PM
34.2 – top layer
23.0 – air directly above   wet bulb – 18.6
19.5 – " breast high   dry " – 22.0

2" in soil where several beetles were – 26.2°C
(4" " " – 21.4)

Came at 11:00 – many beetles active then. Most prob. at noon & 1:00. Later fewer or most caught!

2 species: caught 20 of smaller (C. senilis) & 4 of larger (C. intersignata)

The smaller more definitely to salt flats, the larger more to edge with a little sand. In ground on sand flat at noon were the smaller – plenty about at same time.

2:00 PM
31.8°C – top layer
23.5 – air above (just above gr)
18.8 – " breast high

Observation:

Put 3 cicindelas in terrarium with 3 lizards (2 Uta stansb. and 1 Sceloporus occid.). As each lizard aroused in the first morning it immediately attacked the tiger beetles – pursued, catching + violent shaking – then letting go. (The beetles lost a leg) – then leaving the beetles alone. Each lizard had I work at it. Then from then on they just palgo peaceably. They were just annoyed, sometimes if they crawled over back. After this first encounter the lizards were tempted with mealworms. The Sceloporus ate 3, the others 2 and 1.

March 2, 1968. Preliminary data and field observations concerning the thermal biology of tiger beetles, made while I was "fishing" for a suitable doctoral project at the University of California, Los Angeles. I eventually abandoned this investigation and went back out into the field to search for something else.

1973

PM June 21 - Arrived in Dryden
Rains most of the time till July 4

Fair weather at least during part
of almost every day till July 31.

Upon arrival - see only Bombus queens
Near first g July - see first workers

Now (July 31) see the drones.
Have seen isolated drones of
B. vagans, terricola, fervidus
perplexus + fernansis. Those of
terricola appear to be more common.
These drones foraging for nectar
from Spiraea latifolia.

July 31. Open nest of B. fervidus in
Ⓧ our meadow. Nest is on surface of ground
in mouse nest. Covered with grass— no wax.
The bees are very aggressive. Catch them one
by one as they come out of nest — ~ 40-50.
The only one not flying out of exposed nest
is the queen. She does most of the
buzzing — and incubates even after nest
is exposed. Comb has ~45 pupae. 7 batches (clumps)
of larvae and/or eggs. Honey— less than 2 cells full!

Summer 1973. One of only a few notebook entries when I was at the University of California, Berkeley, and had returned home to Maine in the summer to work on bumblebees in the field. (Scientific notes were separate.)

Migration of Red Admiral, *Vanessa* atalanta rubria

May 11, 1985   Richmond, Vt.

In the afternoon from ~ 2:30 PM → 4:30 PM as I jog along an 18 mile loop I count 512 of the butterflies crossing road in front of me. All but 5 of these are flying in a generally Northeasterly direction, and very fast and straight.

At 5:00 PM I take compass readings of butterflies flying over a plowed field, where they funnel onto it through a valley. I can see them, t take a bearing, for at least 50 paces – 250 feet. All 22 observed are flying in NE direction – with the following degrees to East: 30°, 30°, 25°, 30°, 25, 30, 45, 45, 35, 30, 30, 40, 40/35, 30, 30, 25, 20, 30, 20, 30. $\overline{X}$ = 31.6° (N = 22). (At 6:00 PM – activity has almost stopped). Breeze is very slight, from NNW.

— migr. distrib.

again many – UT summer of 2001!

P.A. Opler + G.O. Krizek in "Butterflies East of the Great Plains" 1984
Vanessa atalanta
1. "Has weak 2-way migration"
2. "Cannot survive harsh winters ~ many parts of northern U.S. ~ canada must be recolonized by migrants each spring!"
3. "Thousands were seen in Florida in April and May, 1953 + 1955, while a great migratory flight was observed in Maine during June, 1957."

May 13 – Hot day – 80°F – none seen

May – 19 – See many *V. atalanta* on leather leaf in Well, Me,

May 11, 1985. Notes on the orientation of the red admiral butterfly (*Vanessa atalanta*).

The same strategy of note-taking then took on a life of its own in my scientific work. As I have searched for answers to biological questions that popped into my mind as I was watching birds or insects, I have meticulously documented my observations, and this documentation has made the difference between simply being a witness to nature and being one who identifies themes and questions.

For me, taking extensive field notes on nature has always been like a free-flowing hunting experience where I was searching for interesting prey. These notes range all the way from ideas that entered my mind at random moments to useful data, and sometimes progressed from one to the other. On one October day in 1984 at my camp in Maine, I heard ravens making a novel call on a distant ridge where I had not seen them before. I was quickly off into the woods on a scientific game trail. Were these birds recruiting others to a kill? This thought motivated me as I hiked through the half-mile of forest and onto the ridge where I heard the calls. When I approached the site I found over a dozen ravens calling and picking at the remains of a dead moose that was mostly covered with brush. I had no idea why these birds would be sharing. They were obviously not a family group, since I knew the local pair, who are territorial and whose young leave by late summer. I knew then that I was onto big and exciting scientific prey. I tried to get my head around the bizarre behavior of these birds, focusing my attention on the bare facts, but my thoughts swirled and I realized I knew nothing. Almost anything might be relevant, so I had to write down everything. Over the following weeks I couldn't pass up any observation or thought, and a notebook on ravens evolved. Protocol was out the window since I had no idea where I was headed or what might be trivial and what important. These notes grew in detail and complexity; the wide trunk of observations divided into many branches, each containing specific questions and hypotheses. These notes of active observation, supplemented by deliberate experimental scenarios to provide more than just random or chance observations, became almost by themselves a book, *Ravens in Winter*. The fruit of the subsequent research by myself and colleagues resulted in thirty-one papers in peer-reviewed journals and then a summary book of the science, *Mind of the Raven*. This raven story matured to the point of predictions (if not theory). Taking notes has always helped me zero in on the interesting questions. They have made the difference between simply observing and being able to get to the meat of the science. When I am in the field collecting information, I am on the lookout for the nascent, the new, and the unexpected that may spring out of the familiar.

After so many years of making observations, there is hardly a thing I encounter that does not connect me in one way or another to familiar ideas or observations. However, I am most interested in the seemingly anomalous. In taking field notes, the way to find these peculiarities is to keep track of many observations that may not appear at the time to be relevant at all. Similar to the way a subtle twist in a blade of grass may betray the presence of game, a single observation in my field notes may stand out against a backdrop of sentences standing in an ordered array. The way that I keep a journal now reflects the chaotic nature of this type of chase. I cannot afford the luxury of presorting data. I don't walk around with a notebook. But I often carry a piece of folded paper in my shirt or pants pocket, along with a pencil stub. The information flow as I jog down our driveway and up our country road may be infinite, and I cannot stop every few feet and record everything. I simply remember most observations while I jog, though I may still record mundane things that catch my eye and that might be useful in identifying something interesting. At these times I'm not trying to solve a problem; instead, I'm open to signs of one.

One year on my annual November deer hunt in the Maine woods while I was perched up in a spruce tree, waiting and watching and shivering in the cold near dusk, I saw several kinglets arrive, and they converged (luckily for me!) at one spot in a nearby thickly branched spruce. I presumed they were responding to the cold by huddling up to spend the night. This observation was the blade out of place, however, because all other birds I had seen at dusk separated, and some individuals entered tree holes. The thoughts that spiraled from witnessing this event resulted in the notes I took. Further observations sharpened my senses until finally, years later, I saw another group of kinglets that I then followed and tracked into a smaller, more climbable tree. I returned there at night, found a "four-pack" of them on a branch, and photographed them huddled into their ball. It was probably the first time this had been recorded or even seen in the wild. The initial kinglet data that aroused my admiration for these birds' ability to survive in extreme cold led to a study of the comparative biology of other organisms, which then culminated in putting it all into a larger framework in *The Winter World*. The experiences of writing this book and hunting deer in November in Maine (mostly unsuccessfully with regard to venison), and my own semi-hibernating experiences at that time, then induced me to consider possible Neanderthal survival strategies from a zoological perspective, which I discuss in *The Summer World*.

Sign is often plentiful, but suitable game is nevertheless often hard to come by. I recall preparing to teach a field ecology course at the University of Minnesota Lake Itasca field station one summer in the

29 Aug. '94

First class tomorrow.

Today I suddenly had the idea to take off, flat-footed, and run my 18 mile Richmond loop — it was almost a dare — this, because I haven't run 18 miles = 2 months. But I _did_ it! Sometimes I wonder if this isn't like shock-treatment that keeps me motivated and alive. Somewhere I read where a poison (from a S. American frog) that normally k_ill_s people, is easily countered if the victim plunges = cold water. The shock affects the nervous system, and it "works"? I've heard the same for treatment of drug overdose (Heroin?). The victim is effectively "dead" — but can sometimes be brought to life, if thrown into a bathtub — only here = addition to the water, add a plugged-in lamp! This yields firewalks, too.

On the run I saw surprisingly _few_ monarch + their caterpillars. Will there be a recent hatch? I picked up or brought back :) 3 caterpillars 2) one Grateful Dead tape 🎵

July 29, 1994. I was starting to train for an ultramarathon race, and using the opportunity to keep track of nature along the way. This notebook is actually a hardcover copy of *The Writings of Albert Einstein, Vol. I*, that was bound with its pages completely blank—a gift from an editor. Inside, I filled every blank page (front and back) with my own field notes.

1970s. Needing to find interesting problems for the students, I was thinking of one of my loves, caterpillars, as a potential example to demonstrate how camouflage and protective resemblance can be a strategy for reducing bird predation. I walked around in the woods

⑥

While I was writing, I heard the flight-calls of geese. They were leaving! I rushed out the door — and heard them coming up over the woods. I stood by the door — they circled over the house, all twelve geese. I called up "Peep — peep — " and the lead goose turned, to come back, leading the whole group behind her. Then the incredible happened. They passed over, and then circled again, and this time she set her wings, started gliding, and extended her feet. She was going to land, as she was coming directly toward me! The small yard is now totally surrounded by densely-leafed-out trees. It's not a place where geese would normally land. It's a hole in the forest. Yet she came on, ~~for~~ a few feet over my head, and then almost crashing into the trees on the other side of the garden. She fluttered to try to break her momentum, and a feather fell out of her wing, drifting down to the ground. She veered sharply, and regained altitude, all eleven behind her, and then she left.

The feather twirled to the ground, and I picked it up. I never thought I could cry over a goose. But I did. I choked up. This was just too incredible to be real. This was too incredible to write about. Things like this ~~just~~ don't just happen. It's more like a fiction. Truth stranger than fiction. More wonderful. Richer. It has been an experience of a lifetime, and I must write it all up to share, even if many will undoubtedly say I am seeing things. But this was Peep. There was no doubt. I'd been writing all summer on the geese. Now this is the ending. The only problem is — it's too good. Nobody will believe me. They think I've made it up. But boy, she did come by last fall, too.

P.S. Went down afterwards — The corn had not been touched.

Heard a catbird "meow". A flock of grackle flew over, making a rushing sound in the air — going east. A blue jay. No fresh beaver sign now. Blue jays + crows.

September 1, 2002. In a notebook concentrating on geese, I entered daily observations at a beaver bog near our house. Eventually these sparked the idea of writing a book, at which point they became more detailed and systematic.

and noticed a fresh basswood leaf lying on the ground. It was partially eaten by a caterpillar but also had a shortened petiole—the tough part of the leaf that caterpillars normally do not eat. Leaf petioles are rigid; they never break. This one must have been laboriously chewed through. I "smelled" a sign of scientific prey; I had previously hunted caterpillars almost exclusively by looking for feeding damage on leaves, so why wouldn't birds do this as well? Had the caterpillar discarded a partially eaten leaf to erase its feeding "tracks" as a predator-avoidance behavior? My random observation of the discarded fresh leaf (when I had just previously studied feeding efficiency in bees) is still one of my favorite scientific trophies because it led to a counterintuitive discovery that would otherwise not have been predicted. Although many spiny and poisonous caterpillars leave leaf tatters and partially eaten leaves on trees, those that are prime bird prey hide their feeding damage by paring down leaves, leaving the scene of their feeding after a meal, or clipping off their partially eaten leaves. In turn, these caterpillar behaviors predicted sensory and cognitive capabilities and the hunting behavior of their predators, mainly birds. Indeed, later tests in the laboratory and an aviary that mimicked the wild confirmed these hypotheses. Even now as I jog along our dirt road my eyes are alert to such sign. I am still finding new "prey" or problems that become journal entries, and then possible topics for future research and subject matter for books.

My most recent example that illustrates this approach concerns nest hygiene behavior in phoebes. One day I listened to my neighbor complain that the motorcycle he had parked under a phoebe nest in his woodshed was covered with guano. Our phoebe, on the other hand, had just fledged its young, and I didn't see a speck of guano under that nest. Weird, I thought, and since it was on my mind I wrote it down. For years I had watched phoebes construct nests on houses, sheds, and barns in Maine and Vermont, and I'm sure I must have seen these birds do a lot of interesting things, many of which I didn't record because they seemed self-evident. I say that because after I started taking notes on the contrasts between these neighboring phoebes, I found things that I had never seen before. I kept coming back to this anomaly of the guano, and wondered why my neighbor's phoebes were messy while mine were neat. Nest hygiene is important for small birds as they hide their nests from predators, but the following year I noticed that the housekeeping in our phoebe nest had become lax and there was an abundance of fe-cal pellets under it after the young fledged. I went back and thumbed through old field notes on a nest I watched for an article I wrote for *Natural History* years earlier, and in those notes I had actually mentioned the feces. I had even counted the pellets to drive home the point that they were more than just incidental. At that point I was convinced I was onto something real: phoebes sometimes meticulously remove what would otherwise be huge amounts of guano, and at other times they don't bother. Just as in the running and raven stories, I had seen sign that led to hours of puzzling, observing and note-taking, and I have now learned that when our phoebes raise their first brood they usually leave the nest area immaculately clean. Only a week later they start to raise a second brood (the last of the year) in the same nest, and for these young they practice nest hygiene only in their early growth stages. These obser-vations raise many new and interesting questions. Do the birds leave the fecal pellets of their second brood because they don't need to put energy into nest hygiene since the nest won't be used again and predators no longer need to be kept away? What happens when or where the birds raise only one brood? Do the parents somehow "know" that nest hygiene under the nest is no longer necessary when it will not be used anymore? Is the nest hygiene related to nutrient recycling and hence diet that might change seasonally? Is lack of fecal removal also observed in other cave nesters, such as barn swallows, whose nests are well protected? At this point in my process I have amassed enough observations and questions relating to this issue that it is worth pursuing as a scientific study. If I continue, my notes will now have to become more precise and systematic, and they will be destined for a file dedicated to the project. I do this to keep all information and data about this project together and separate from other observations, because my field notebooks contain a lot of unordered information.

As a scientific naturalist, I gain my inspiration from overlaying my observations onto previous knowledge in my notebooks. At a glance, my journal seems to be a mess. It is not meant to be seen or read, except by me, and often not even that. When I write in it, it is usually in haste. I use any stray implement at hand. I have no system, no object or goal in mind. The notebook allows for spontaneity, a counterbalance to my ideal of orderly scientific objectivity. It is my wild side that explores without restraint, without inhibitions. Its value is usually derived less from its contents and more from the exercise of writing things down that forces me to pay attention and to remember. This process slows my thinking and serves as a first crude filter for the natural breeze of data that passes by in a continual stream.

I seldom adhere to a formula, pattern, subject, or format for taking my field notes. There has, however, been a rough evolution from reasonably focused and sparse notes to a haphazard mélange. I now scribble on my folded piece of paper anything I find interesting or relevant. Later, when I'm at home and sitting down, I will pull it out of my pocket and consider the crude notes. I then write in my notebook, which is now usually a spiral notebook, 8.5 × 11 inch lined, such as I used for taking class notes in college. Since I usually don't know what category to put my entry into, but may want to retrieve it later, I paginate and sometimes underline key words or topics for orientation and easier reference later.

I've been keeping journals of one sort or another since I was a teenager, and if there is one thing I can now confidently say about all this scribbling and note-taking, it is that if it wasn't written down, it didn't happen. The more I wrote the more that did happen, because this process stirs up ideas. Stopping to sit and write costs time and energy, and some biologists feel that it should be discouraged. A very prominent one (Ernst Mayr) suggested to me that writing in a notebook is a "waste of time" because it is distracting. In his case it probably was. He was mainly interested in one theme. Also, perhaps he had an eidetic memory, and therefore is an example of someone who did not need to take notes (because he ostensibly remembered everything). I am different in temperament and have found taking notes an enjoyable and useful tool that has helped me concentrate or at least pay attention, and to extract the signal from the overwhelming noise of nature.

As I write this I am at my camp in the Maine woods. The local raven pair that I know from my previous studies in the aviary is making a nest in the nearby pines. They offer their calls as pleasant and satisfying background noise. I pay little attention to them as I watch crane flies doing a mating dance on the outhouse. I look over my notes that I took just a few days ago, and after reading them take more detailed notes on the crane flies' activities and make a sketch as well, because I'm thinking of including them in a book chapter on flies. I mix in observations

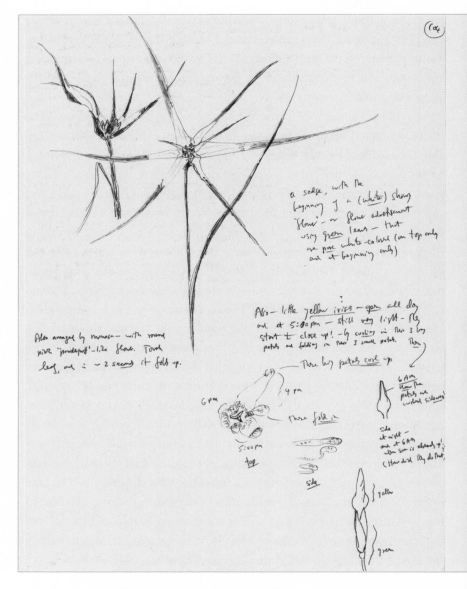

May 26, 2006. Entries made during a trip to Puerto Rico with my son Stuart to show him the tropics. In the first, the "demonstration" of apparent flower evolution from leaves, and plant movement of both leaves and flower petals, caught my eye. The second entry contains notes on caterpillar feeding and leaf damage.

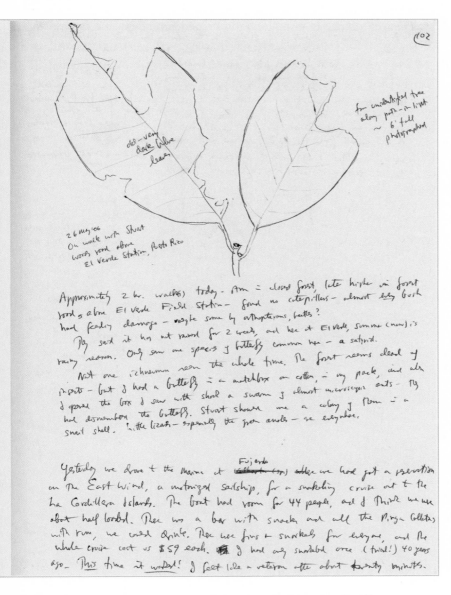

of the weather with those on our tame Canada geese, and reflect on things written or to be written. I consider the crates of notes I have, spanning exactly half a century and starting from my small running notebooks, and how all this writing has allowed me to focus on captivating questions. Revisiting my notes on the wash of green colors that fascinated me during a recent early spring walk one morning in Huckleberry Bog, a study area where I used to follow bumblebees, makes me realize how note-taking helped transform me from a young boy on barefoot runs who passively observed the tangled bank of the Maine woods into a naturalist-scientist who is an active participant in unraveling the mysteries of the natural world.

3

One and a Half Cheers
for List-Keeping

KENN KAUFMAN

IN SERIOUS STUDIES of bird distribution, an annotated list of species detected is at the heart of an effective set of field notes. In recreational birding, a list of species detected is at the heart of one of the most frivolous games ever devised. This shared element has led to persistent confusion, at least among birders, between the scientific act of keeping field notes and the game of list-chasing. That in itself is reason enough for a discussion of listing to be included in this volume. But beyond that, the phenomenon of list-keeping is worth examining in its own right, for the influences—both positive and negative—that it can have on the development of a naturalist, and on the advancement of the natural sciences. I have been involved with birds and other areas of natural history for years, and have alternated between being a rabid lister and being at least mildly antilisting, so I can talk about the subject from both sides.

If both approaches, note-taking and list-chasing, may produce a species list of some kind, what's the difference? As in so many other realms of human endeavor, the difference lies in the intention. In a bird survey or census, the area to be covered and usually the time period allowed will be established ahead of time, and focus will be placed on counting species and individuals in some standardized way so that the effort can be duplicated and the results compared. In listing, the area and the time period may also be predetermined, but the focus will be on finding just as many species as possible.

The similarities and differences might be illustrated most clearly by comparing two kinds of road-based counts, the Breeding Bird Survey

2 DECEMBER 1992 — SOUTH ORKNEYS
— WE HAD PLANNED TO REACH CORONATION
ISLAND EARLY IN THE MORNING, BUT WERE
SLOWED DOWN BY ICE & FOG, SO DIDN'T
ARRIVE UNTIL MID-DAY. PICKED UP SOME
BRITISH ANTARCTIC SURVEY PEOPLE FROM
THEIR BASE ON SIGNY ISLAND & THEN
PUT ASHORE AT SHINGLE COVE, CORONATION I.,
ABOUT 1430 – 1645. DURING THE AFTERNOON
THE WEATHER WAS INCREDIBLE, W/ CLEAR
SKY, NO WIND, TEMP ABOUT 40°F.

    SP. REL. (AT SEA / VIC CORONATION)
BLACK-BROWED ALBATROSS 1/ —
S. GIANT PETREL 15/10
SOUTHERN FULMAR 10/ —
CAPE PETREL 1000/100+ — NESTING ON LEDGES
SNOW PETREL 40+/20+ — MOST APPARENTLY
    GOING TO NEST ABOUT HIGH CLIFFS ABOVE
    SHINGLE COVE
ANTARCTIC PRION 30/20 — ALSO FOUND
    MANY DETACHED WINGS ON THE ISLAND;
    APPARENTLY THE SKUAS PREY HEAVILY ON
    THE PRIONS. SAW ONE ON A NEST, IN
    A DEEP CREVICE UNDER A ROCK.
WILSON'S STORM-PETREL 35/10
BLACK-BELLIED STORM-PETREL 2/ —
GENTOO PENGUIN 4/ —
CHINSTRAP PENGUIN 35/ —
ADELIE PENGUIN 50/1000 — MOST IN
    COLONY STILL ON EGGS, BUT SOME
    HAD NEWLY HATCHED YOUNG
BLUE-EYED CORMORANT (P.A. BRANSFIELDENSIS)
    5/35

A page from my notes on a day in the South Orkney Islands in the Antarctic region. When I travel as a lecturer and guide on nature tours, my time is limited so my notes tend to be brief, but daily notes almost always include a list of species. Here I used a slash to separate numbers of individuals seen at sea from those seen in the vicinity of Coronation Island.

and the Big Day. On the Breeding Bird Survey, the route is established for the participants ahead of time by the U.S. Fish & Wildlife Service or the Canadian Wildlife Service. Starting half an hour before local dawn, we make fifty stops, spaced half a mile apart along a 24.5-mile route, and at each stop we spend exactly three minutes counting every individual bird of every species seen or heard within a quarter-mile radius. By contrast, on a Big Day, we have planned the route ahead of time, although we're usually revising it up to the last minute. Starting early—often at midnight—we make a series of stops along a route that may stretch to hundreds of miles. The time spent at each stop is variable, but most stops, especially in daylight, are short and frantic. We don't count individuals—all we need is one of each species, and after the first one for the day, we tune out the others of its kind. The point is to find as many species as possible before the following midnight.

A casual observer might not see much difference between these two activities. But the Breeding Bird Survey, conducted over thousands of routes over several decades, produces our best index of population trends for many North American bird species. The Big Day produces nothing but bragging rights. Both pursuits are enjoyable in their own way, but the value of the first is far more evident.

Of course there are many variations and many shades of gray between these extremes. It's even possible to take either approach with some kinds of field activities, such as the Christmas Bird Count (CBC). Established around 1900, before the era of standardized census protocols, the CBC has a very loose methodology. Observers divide into small field parties and spend one calendar day covering choice areas within a fifteen-mile-diameter circle, counting every bird they can identify. Even within this loose framework, some observers take a conscientious approach, steadfastly covering the same area every year in the same way. That's the best way to do it, of course, but I confess that I find it hard to stick to this sober census. Instead, I find myself falling back on the CBC approach that my friends and I always used as teenagers. We would race around our assigned area like maniacs, trying to ferret out every species we could, and trying to finish before dark so we could go "poach" in some other party's area, to see if we could find some rarity that the other guys had missed. We did keep track of individuals, in a rough way, but the emphasis was on finding more species than the next party, pushing

15 OCTOBER 2005 — CAPE HATTERAS POINT — STAYED THE NIGHT OF THE 14TH AT BREAKWATER INN, NEXT TO ODEN'S DOCK IN HATTERAS. OUR PELAGIC TRIP FOR TODAY WAS CANCELLED BECAUSE OF HEAVY SEAS SO WE HAD TODAY TO EXPLORE. THERE HAD BEEN BAD WEATHER FOR SEVERAL DAYS PRIOR (RAIN, INCL VERY HEAVY RAIN 5-6 DAYS BEFORE, STRONG NORTHERLY WINDS, TO 30 MPH) BUT TODAY DAWNED TOTALLY CLEAR, + WITH NORTH WINDS OF 10-15 MPH THAT DECREASED TO ALMOST NOTHING BY LATE AFTERNOON. IN THE MORNING WE LOOKED AT SOME OTHER SPOTS INCL. THE "NATIVE AMERICAN MUSEUM" IN FRISCO (FUNKY LITTLE PLACE W/ A MINOR NATURE TRAIL OUT BACK) BUT OUR MOST INTERESTING STOP WAS AT CAPE HATTERAS POINT, ABOUT 1330 - 1800 — PARKED NEAR SOUTHERN END OF PARKED ROAD + WALKED OUT SOUTHWARD THROUGH DUNES, ALONG BEACH, AROUND SALT POINT, + AROUND DESIGNATED TERN NESTING AREA NEAR TIP OF POINT. HIGHLIGHTS HERE WERE NUMBERS OF PEREGRINES (OFTEN 4 VISIBLE AT ONCE, SOMETIMES 5 OR 6), NUMBERS OF LARGE GULLS + TERNS (& ABSENCE OF SMALL ONES), + PRESENCE OF ODD ASSORTMENT OF LANDBIRD MIGRANTS.

Two pages out of the middle of my notes from a day on the Outer Banks of North Carolina. The first page was occupied with detailed notes on the day, and the pages beyond those simply continued the species list, following the sequence of the American Ornithologists' Union checklist of North American birds. A list in this format, written at the end of the day, allows for extended comments where warranted, as with the Peregrine Falcons.

the count total higher than last year's or higher than the tally on that other count downstate. The species total was the thing.

It's easy to see why listing, as a game, developed with birds before it was applied to other groups of organisms. It satisfies the collecting urge (and the collecting of bird specimens, as a hobby, became illegal for most people a century ago). Birds provide enough variety to keep things interesting: over most of North America, except for a few areas of the

15 OCT 2005 AT CAPE HATTERAS POINT, NC
CONTINUED
(PEREGRINE CONTINUED) BOTH LOW OVER THE
WATER & HIGH OVERHEAD — ALSO SOME INTERACTION
AMONG PEREGRINES, & SAW ONE MAKE A PASS AT
A VULTURE. SAW A COUPLE AT BUXTON & FRISCO
EARLIER IN THE DAY. OUT AT THE POINT THEY
WERE MOST EVIDENT BETWEEN 1400 & 1630 BUT
WE SAW A COUPLE LATER.
AM. KESTREL — A COUPLE GOING BY OUT AT
POINT; THEY DIDN'T STICK AROUND.
MERLIN — SAW NONE OUT AT POINT. BUT ON
WAY BACK AROUND 1800 SAW 2 NR LIGHTHOUSE.
AM. OYSTERCATCHER — 2 AT POINT
BLACK-BELLIED PLOVER — 1 HEARD, 1 NEAR BEACH
WILLET — A COUPLE OF FLOCKS OF 10 - 20
GR. YELLOWLEGS - 3 FLEW OVER TOGETHER
LESSER YELLOWLEGS - 2 SINGLE FLYBYS
SANDERLING — 2 - 3 LARGE FLOCKS (35+)
    FLYING AROUND & AT WATER'S EDGE
DUNLIN — 3 WITH ONE SANDERLING FLOCK
BAIRD'S SANDP. — ONE LONE BIRD
RUDDY TURNSTONE — 3
GR BLACK-B GULL — 200+ — MOSTLY RESTING
WITHIN AREA DESIGNATED AS TERN NESTING AREA;
PROB. 2/3 ADULTS

high Arctic, there are hundreds of bird species to be found, and a bird list in excess of one hundred in a day would be possible in most places. By contrast, in most of North America, a daily mammal list of much more than a dozen species would be surprising. (On trips to East Africa, of course, friends and I have delighted in seeing how many mammal species we could list in a day.) In addition, almost all birds can be identified readily in the field. There are other groups of organisms that present much higher diversity but also much more extreme challenges in identification. For example, in North America, the number of known fly species outnumbers the number of bird species twenty to one, but identifying most of those flies to species would be an onerous chore even for a devoted student of Diptera. Hobbyists are not likely to go there anytime soon.

Still, the listing game is being applied to some life forms besides birds. Now that close-focusing binoculars and more field guides have improved the odds for butterfly and dragonfly identification, more observers have begun trying for big lists for these groups. Plant listing games, too, have cropped up here and there. Activities of this sort are likely in the future as resources for identification become more available, so a consideration of the pros and cons of listing is not just a subject for ornithologists.

The negatives of an extreme list-chasing approach are probably evident to anyone reading this volume. Dashing around a county or state for a day, or around a continent or the world for a year, in simple pursuit of a high species total is not the most productive thing to be doing, regardless of how much fun it is (and I know from personal experience that it can be a *lot* of fun).

Beyond that, a continued emphasis on just finding and checking off new species can eventually stunt a person's development as a naturalist and limit her ability to learn more. I worked for several years as a leader of birding tours, and I met a few sad individuals who were so focused on adding to their life lists that they would refuse to look at a bird species that they had seen before, no matter how spectacular the view or how fascinating its behavior of the moment might be. "I don't need that bird" was their standard reply. I knew one man who went repeatedly on organized spring birding trips to St. Lawrence Island, Alaska, hoping to see strays from Siberia to add to his North American life list. He would sit in the lodge by the radio, waiting for one of the other birders to call in and report some rarity that would be new for him. One year there was nothing found that he hadn't already seen before, so as far as we could tell, he didn't look at a bird the whole time he was there. And this at a place where the "everyday" birds, utterly different from those in most of the United States, occur in mind-blowing concentrations. Such an approach

PASSERIFORMES (Troglodytidae)
Species	Code
Carolina Wren	o D,M
House Wren	r D,M
Winter Wren	c E
Marsh Wren	u M

PASSERIFORMES (Regulidae)
Species	Code
Golden-crowned Kinglet	c E
Ruby-crowned Kinglet	u E

PASSERIFORMES (Sylviidae)
Species	Code
Blue-gray Gnatcatcher	r D,M

PASSERIFORMES (Turdidae)
Species	Code
Eastern Bluebird	o M
Veery	u D
Swainson's Thrush	c E
Hermit Thrush	c E
Wood Thrush	u D
American Robin	c D,M

PASSERIFORMES (Mimidae)
Species	Code
Gray Catbird	c D,M
Northern Mockingbird	r D,M
Brown Thrasher	u D,M

PASSERIFORMES (Sturnidae)
Species	Code
European Starling	c D,M

PASSERIFORMES (Bombycillidae)
Species	Code
Cedar Waxwing	c D,E

PASSERIFORMES (Parulidae)
Species	Code
Blue-winged Warbler	r M
Tennessee Warbler	o D,E
Orange-crowned Warbler	r M
Nashville Warbler	o E
Northern Parula	c D
Yellow Warbler	c D,M
Chestnut-sided Warbler	u M
Magnolia Warbler	c E
Cape May Warbler	o E
Black-throated Blue Warbler	u D,E
Yellow-rumped Warbler	c E
Black-throated Green Warbler	c D,E
Blackburnian Warbler	u E
Pine Warbler	u E
Prairie Warbler	o E
Bay-breasted Warbler	o D,E
Blackpoll Warbler	o E
Black-and-white Warbler	u M
American Redstart	u D,M
Ovenbird	u D
Northern Waterthrush	o E,F
Mourning Warbler	r F,M
Common Yellowthroat	c F,M
Wilson's Warbler	r D
Canada Warbler	u D,F
Yellow-breasted Chat	r M

PASSERIFORMES (Thraupidae)
Species	Code
Scarlet Tanager	u D

PASSERIFORMES (Emberizidae)
Species	Code
Eastern Towhee	u D,M
Chipping Sparrow	c M
Savannah Sparrow	o I,M
Nelson's Sharp-tailed Sparrow	u I
Saltmarsh Sharp-tailed Sparrow	r I
Seaside Sparrow	r I
Song Sparrow	c D,M
Lincoln's Sparrow	o M
Swamp Sparrow	c F
White-throated Sparrow	c E,M
White-crowned Sparrow	r M
Dark-eyed Junco	c E

PASSERIFORMES (Cardinalidae)
Species	Code
Northern Cardinal	o M
Rose-breasted Grosbeak	u D
Indigo Bunting	o M
Dickcissel	r D,M

PASSERIFORMES (Icteridae)
Species	Code
Bobolink	u M
Red-winged Blackbird	c F,M
Eastern Meadowlark	u M
Common Grackle	u M
Brown-headed Cowbird	c E,M
Orchard Oriole	r D,M
Baltimore Oriole	u D

PASSERIFORMES (Fringillidae)
Species	Code
Purple Finch	u D,E
House Finch	u M
Red Crossbill	u* E
White-winged Crossbill	u* E
Pine Siskin	u E,M
American Goldfinch	c M
Evening Grosbeak	o E

PASSERIFORMES (Passeridae)
Species	Code
House Sparrow (I)	c D,M

Notes:

AUDUBON CAMP IN MAINE

Summer Bird Checklist

a abundant - widespread & easily found in proper habitat
c common - certain to be seen or heard in suitable habitat
u uncommon - present but not certain to be seen or heard
o occasional - seen a few times each season
r rare - seen only a few times in 10 years
* irregular - intermittently common or absent

E Evergreen/coniferous forest
D Deciduous Forest
M Meadow/Thicket
F Fresh water including marshes
I Inshore
P Pelagic (ocean-going)

Date/Location
A _____
B _____
C _____
D _____
E _____

GAVIIFORMES (Gaviidae)
Species	A B C D E	Code
Red-throated Loon		r I
Common Loon		c I

PODICIPEDIFORMES (Podicipedidae)
Species	Code
Pied-billed Grebe	o F
Horned Grebe	r I
Red-necked Grebe	r I

PROCELLARIIFORMES (Procellariidae)
Species	Code
Greater Shearwater (N)	r P
Sooty Shearwater (N)	r P
Manx Shearwater	r P

PROCELLARIIFORMES (Hydrobatidae)
Species	Code
Wilson's Storm-Petrel (N)	o P
Leach's Storm-Petrel	o P

PELECANIFORMES (Sulidae)
Species	Code
Northern Gannet	o P

Here's an example of a field checklist designed for just one season at one place (the Audubon Camp on Hog Island, Maine, and surrounding areas). In a very compact format, the checklist gives a good idea of what bird species can be expected by participants in the summer educational programs at the camp. Boxes before each species name allow checkmarks for five different days or sites—or numbers of individuals recorded, for those who can write small enough. Courtesy of Maine Audubon.

is unfathomable to most birders, but it represents the listing approach taken to its extreme.

So an exclusive focus on listing as a game or sport can have a less than desirable impact on the person playing the game and can hardly be better than neutral or irrelevant for the community at large. There are, however, less obvious ways in which the game of listing can be beneficial for naturalists or budding field biologists, and can even make a contribution to science.

I've had the opportunity to interact with a lot of beginning naturalists, but the experience I watched most closely was my own. For me, a general fascination with animals crystallized into a focus on birds when I was six years old. I didn't know another soul who was interested in birds, so I learned on my own from books, and learned at a slow pace, gradually figuring out the birdlife of our Indiana suburbs. In some book I ran across the idea that I should be keeping a life list of all the species I had identified, so at the age of about eight I started one, writing down my birds in a notebook and watching with pride as my life list passed twenty species and headed toward twenty-five.

It would make a good story if I could tell you that I still have that notebook and that my life list now stands at some specific number. But in fact the original notebook was lost and replaced and started over a dozen times before I was twelve. Although I was a rabid list-chaser throughout my teen years, my interest in maintaining a life list faded in my early twenties. Today I don't have the foggiest idea how many bird species I've encountered. No such list is tallied anywhere. I could sit down and compile such a list from my detailed daily notes from scores of trips all over the world, but that would take weeks, and I just don't care.

Having said that, however, I still think that the list-keeping compulsion of my early years was valuable. In fact, I would say that for a person just getting started in some area of natural history, an unabashed focus on list-chasing is a good thing, at least for a while. The trick is knowing when to stop. But you can always worry about that later.

Listing, as a game or sport, works best in a setting where the fauna and flora are already fairly well known. Biological explorers venturing into brand-new turf will be eagerly seeking novelty, of course, but they are operating under a different mindset. To play the listing game in unexplored territory would be like playing golf without a golf course: you could hit the ball all day, but you wouldn't have any way to measure how well you were doing. A well-defined set of available species establishes the course, or at least establishes the limits of the playing field. If the number of species to be found were infinite, counting species would become like counting raindrops, and no number would ever mean anything.

Page 8

	Southern	Northern
ng-billed Gull	M,W-lc	M,W-u
alifornia Gull	M,W-lu	M,W-u
erring Gull	M,W-r	M,W-ca
hayer's Gull	W-acc	FM-acc ?
Western Gull	SpV-acc	
laucous-winged Gull	W, SuV-acc	
laucous Gull	W-acc	
abine's Gull	(F)M-r	FM-r
ack-legged Kittiwake	(F)M,W-ca	FM-acc
ull-billed Tern	SpV-acc	
aspian Tern	(F)M-lc; W-lr	M-ca
egant Tern	Sp, SuV-acc	
ommon Tern	(F)M-lu	(F)M-r
rctic Tern	(F)M-acc	FM-acc
orster's Tern	(F)M-lc; W-ilr	M-u
east Tern	(Sp)M-r	Su-acc
lack Tern	(F)M-c	M-u
lack Skimmer	(Su)V-ca	

Columbidae - Pigeons and Doves

	Southern	Northern
ock Dove, n (int)	R-c	R-c
and-tailed Pigeon, n	Su-c; M,W-ir	Su-c; M-r; W-acc
urasian Collared-Dove, n (int)	V-r	V-ca
White-winged Dove, n	Su-c; W-r	Su-r; W-ca
ourning Dove, n	P-c	Su,M-c; W-u
ca Dove, n	R-c	(F)V-ca
ommon Ground-Dove, n	Su-lc; W-u	FV-ca
uddy Ground-Dove, n?	F,WV-r; SuV-acc	

Psittacidae - Parrots

	Southern	Northern
hick-billed Parrot	(W)V-ca	

Cuculidae - Cuckoos, Roadrunners, and Anis

	Southern	Northern
lack-billed Cuckoo	FM-acc	
ellow-billed Cuckoo, n	Su-lc; M-r; W-acc	Su-lu
reater Roadrunner, n	R-c	R-u
roove-billed Ani	(Su,F)V-ca	(F)V-acc

Tytonidae - Barn Owls

	Southern	Northern
arn Owl, n	P-u	Su-lr

Strigidae - Typical Owls

	Southern	Northern
lammulated Owl, n	Su-c; W-acc	Su-c
Western Screech-Owl, n	R-c	R-lu
Whiskered Screech-Owl, n	R-c	
reat Horned Owl, n	R-c	R-c
orthern Pygmy-Owl		
californicum form, n	R-lr	R-lu
gnoma form, n		R-u
erruginous Pygmy-Owl, n	R-lr	
lf Owl, n	Su-c	
urrowing Owl, n	R-lu	Su,M-lu; W-acc
potted Owl, n	R-u	R-lu
ong-eared Owl, n	R-ilr; W-r	Su-lr; W-r
hort-eared Owl	W-r	M-ca; Su,W-acc
orthern Saw-whet Owl, n	R-ilr; W-ca	R-lu

Page 9

Caprimulgidae - Nighthawks and Nightjars

	Southern	Northern
Lesser Nighthawk, n	Su-c; W-ca	Su-lr; M-ca
Common Nighthawk, n	Su-lc	Su-c
Common Poorwill, n	Su-c; W-ir	Su-c
Buff-collared Nightjar, n	Su-lr	
Whip-poor-will		
arizonae form, n	Su-c; W-acc	Su-lc
vociferus form	FV-acc	

Apodidae - Swifts

	Southern	Northern
[Black Swift]	Su,M-ca	M-ca
Chimney Swift	Su-ilr; M-ca	
Vaux's Swift	M-u	M-r
White-throated Swift	Su-c; M,W-lc	Su,M-c

Trochilidae - Hummingbirds

	Southern	Northern
Broad-billed Hummingbird, n	Su-c; W-r	
White-eared Hummingbird, n	Su-ilr; W-ca	
Berylline Hummingbird, n	(Su)V-ca	
Cinnamon Hummingbird	SuV-acc	
Violet-crowned Hummingbird, n	Su-lu; V,W-ca	
Blue-throated Hummingbird, n	Su-lu; W-lr	
Magnificent Hummingbird, n	Su-c; W-lr	Su-lr
Plain-capped Starthroat, n	(Su)V-ca	
Lucifer Hummingbird, n	Su-ilr; V-r	
Black-chinned Hummingbird, n	Su,M-c; W-acc	Su,M-u
Anna's Hummingbird, n	(F)M,W-c; Su-lu	(Su,F)M-ca
Costa's Hummingbird, n	W,Sp-c; Su,F-lr	Sp-lu
Calliope Hummingbird	(F)M-u; W-acc	FM-u, SpM-acc
Bumblebee Hummingbird	SuV-acc	
Broad-tailed Hummingbird, n	Su,M-c; W-acc	Su,M-c
Rufous Hummingbird	(F)M-c; W-acc	FM-c
Allen's Hummingbird	(F)M-r	

Trogonidae - Trogons

	Southern	Northern
Elegant Trogon, n	Su-lu; W-lr	
Eared Trogon, n	(Su,F)V-ca	SuV-acc

Alcedinidae - Kingfishers

	Southern	Northern
Belted Kingfisher, n	M,W-u; Su-lr	M,W-u; Su-lr
Green Kingfisher, n	P-lr; (F,W)V-ir	

Picidae - Woodpeckers

	Southern	Northern
Lewis's Woodpecker, n	M,W-r	P-lu; M-u
Red-headed Woodpecker, n	M,W-acc	FV-acc
Acorn Woodpecker, n	R-c	R-lc
Gila Woodpecker, n	R-c	FV-acc
Williamson's Sapsucker, n	M,W-r	R-u; (F)M-r
Yellow-bellied Sapsucker, n	(F)M,W-ca	FM-acc
Red-naped Sapsucker, n	M,W-u; Su-lr?	M-c; W,Su-u
Red-breasted Sapsucker, n	M,W-ca; Su-acc	
Ladder-backed Woodpecker, n	R-c	R-lu
Downy Woodpecker, n	M,W-ca	R-lr; M,W-u
Hairy Woodpecker, n	R-u	R-u
Arizona Woodpecker, n	R-u	
Three-toed Woodpecker, n		R-r

An annotated checklist of the birds (or other organisms) of a specific region can be a powerful learning tool. Even a naturalist who uses the list simply as part of a game, trying to check off more species, is bound to learn things about status and distribution. This annotated list produced by the Arizona Bird Records Committee gives detailed status for southern and northern Arizona and serves as a mini-treatise on bird distribution in the state. Courtesy of Gary Rosenberg and Dave Stejskal.

In my own early adventures, I was fascinated with birds but only luke-warm on keeping my life list until after my family moved from Indiana to Kansas. There I discovered a little checklist card published by the Wichita Audubon Society, listing all the birds likely to be found within fifty miles of Wichita. It was a revelation that caused an immediate shift

On a trip to Europe in 1991 I spent a long day on the coast of northern France looking closely at gulls, with the aim of understanding the field identification of various forms. That day resulted in detailed notes and sketches (here, Yellowlegged, Lesser Blackbacked, and European Herring Gulls), with sketches done on the spot and notes added that evening, but I kept no list of species for the day. My focus on a handful of gulls was so intense that I failed to notice what other birds might have been around.

LARUS
LE PORTEL, BOULOGNE, FRANCE
21 OCT 1991

WINTER
ADULTS

BILL GIVES THE IMPRESSION THAT GONYDEAL ANGLE IS SUBTLY FARTHER BACK THAN ON ARGENTEUS, BUT NOT EXTENDING DOWNWARD AS MUCH,

SO THAT DISTAL PART OF BILL LOOKS EVENLY THICK BUT NOT BULBOUS

FOREHEAD NOT LONG & SLOPING LIKE FUSCUS, NOR AS ANGLED AT REAR OF CROWN — MORE ROUNDED, LOOKS "PUFFY"

EYE SEEMS SMALL RED ORBITAL RING OBVIOUS ON THESE BIRDS (21 OCT)

L. CACHINNANS MICHAHELLIS

SMALL EYE & ROUNDED HEAD MAY BE THE ELEMENTS THAT CREATE THE MORE "GENTLE" LOOK

SLOPING FOREHEAD, ANGULAR HEAD, OFTEN ACCENTUATED BY PATTERN OF DARK MARKINGS BEHIND EYE

L. FUSCUS GRAELLSII

BULGING FOREHEAD MUCH DARKNESS AROUND EYE, BUT WIDE PALE UPPER "EYELID" REMAINS CONSPICUOUS

GONYDEAL ANGLE USUALLY BULGES DOWNWARD, GIVING BILL A LESS EVEN-EDGED APPEARANCE THAN IN OTHER 2 FORMS

L. ARGENTATUS ARGENTEUS

ATTEMPTING TO ANALYZE SHAPE & PATTERN ELEMENTS THAT GIVE EACH OF THESE THEIR CHARACTERISTIC EXPRESSION — BASED ON STUDY OF SEVERAL INDIVIDUALS OF EACH, PLUS SCANNING OF OTHERS in my awareness of birds. How many of these species could I find? Could I find anything that was not on the list?

Without my fully realizing it, that little checklist card became my framework for learning for the next couple of years, and I have since come to believe that a local bird checklist is one of the most powerful learning tools that a new birder or ornithologist can use. In going over my list repeatedly (practically every day), I was reviewing both the birds I had already seen and the ones that still eluded me. The search for the birds I hadn't seen sharpened my awareness as I tried to find out about their habitats and migrations, and I went out repeatedly hoping to find them. The search for new birds was also an incentive to get out in the field more often and to explore different areas at different seasons. It added greatly to the wear and tear on my bicycle and my tennis shoes, and I gained greater familiarity with the common birds while I sought the rare ones.

At the same time that it inspired my investigations, the list was also teaching me the rudiments of classification. Almost all bird checklists (aside from an unfortunate few that use alphabetical order) are arranged taxonomically, the species divided up by families or by similar group-ings, following the sequence of some formal list. In North America, checklists almost always follow the sequence established by the Ameri-can Ornithologists' Union's Committee on Classification and Nomen-clature, and my little checklist from Wichita Audubon was no exception. As I reviewed it day after day, I was noting the fact that I could expect to see three members of the grebe family, and that these came much ear-lier in the sequence than the seven possible members of the woodpecker family or the thirty possible members of the warbler family. Almost un-consciously, I memorized the sequence of the official AOU list.

All of this checklist-induced field time seemed to lead, quite natu-rally, to a curiosity about how many kinds of birds I could find in a day. In addition to exploring new areas, I would ride my bike to the river to tick off the resident chickadees, over to the sandpit ponds to check for lingering ducks, out to the fields by the flood-control channel to look for migrating sparrows. I began comparing my daily totals to what I had been able to find previously at this time of year—could I find sixty spe-cies today? Seventy? And of course I was recording my daily list every

13 MAY 1973   COCHISE + SANTA CRUZ COUNTIES, ARIZONA
TEMP 70-90°F   SKIES MOSTLY CLEAR LIGHT WIND
OBSERVERS: K.K. TED PARKER, MARK ROBBINS, JOEL
GREENBERG, DAVE HAYWARD
IN FIELD 0330-2000

~ AFTER ANOTHER RELATIVELY SLEEPLESS NIGHT, WE STARTED
EARLY FOR SCOTIA CANYON. SEEING HOW MANY MILES OF
COMPLETELY UNTRAVELED DIRT ROADS ARE INVOLVED IN GETTING THERE,
I'M GLAD I DIDN'T TRY TO HITCH IN. SCOTIA CANYON WAS AN
INTERESTING PLACE, OAK-APACHE PINE ASSOCIATION WITH SYCAMORES ALONG
THE STREAM. WE HEARD A BUFF-BREASTED FLYCATCHER ALMOST AS
SOON AS WE GOT OUT OF THE CAR: A TYPICAL EXPLOSIVE
EMPIDONAX SONG, WHICH SOUNDED TO ME LIKE "CHEE-WIP!" WE
QUICKLY TRACKED DOWN A PAIR, WHICH WERE RATHER FLIGHTY AT
FIRST, BUT EVENTUALLY THEY STAYED PUT LONG ENOUGH FOR GOOD
VIEWING. A SHORT TIME LATER WE FOUND ANOTHER PAIR
NEST-BUILDING. AT SCOTIA WE ALSO FOUND AN OLIVACEOUS
FLYCATCHER'S NEST, AND SOME INTERESTING LIZARDS AROUND AN
ABANDONED FARMSTEAD. WE ALSO MET AN OLD MAN THERE, WHO
TOLD US A LITTLE OF THE HISTORY OF THE NEARBY TOWN OF SUNNYSIDE
AND THEN GAVE US EACH A LITTLE BOOKLET OF BIBLE QUOTES
~ WE CONTINUED AROUND THE SOUTH END OF THE HUACHUCAS
AND UP TO RAMSEY CANYON. NOT MUCH WAS HAPPENING AT
RAMSEY. HUMMERS WERE IN RELATIVELY LOW NUMBERS, AND
THE SULPHUR-BELLIES AND TROGONS WEREN'T IN YET. AROUND NOON
WE LEFT RAMSEY. HAD A SHORT BUT BITTER ARGUMENT WITH THE
"UNCLE SAM'S RESTAURANT" IN SIERRA VISTA AND THEN HAD LUNCH
AT SOME HAMBURGER PLACE. CONTINUED TO THE NATURE CONSERVANCY

PLACE AT PATAGONIA, WHERE THE BLACK HAWK
APPEARED ON CUE. WE ALSO HAD A TOWNSEND'S WARBLER
THERE, AND SEVERAL BEARDLESS FLYCATCHERS. AFTER
RELOCATING EVERYONE, AND STAVING OFF DEHYDRATION
WITH A MILKSHAKE IN PATAGONIA, WE WENT DOWN TO THE
ROADSIDE REST. THE BECARD WAS CALLING IN THE COTTONWOODS
ACROSS THE ROAD. STARTING UP THE SPARROW CANYON, WE
CHANCED ON TWO HUMMINGBIRD NESTS LESS THAN 100 YARDS
APART, ONE COSTA'S AND ONE BROAD-BILLED. BY THIS TIME
THE SUN WAS ALREADY DOWN BEHIND THE RIDGE AND NOT
A SINGLE BIRD WAS CALLING IN THE CANYON. IT LOOKED
FOR A WHILE AS IF WE MIGHT MISS THE FIVE-STRIPED
SPARROW, BUT TED AND I MADE A LAST-DITCH ATTEMPT,
GOING DOWN THE WEST SIDE OF THE CANYON AND UP THE
NEXT RIDGE, AND WE FOUND ONE OF THE BIRDS. WE
CALLED THE OTHER GUYS AND WATCHED IT FOR 15 MINUTES.
S - SCOTIA CANYON
SPECIES RECORDED: R - RAMSEY   V - PONDS AT SIERRA VISTA
                     P - SONOITA CREEK

EARED GREBE V 1   WILSON'S PHALAROPE V ?
GR. W. TEAL V 4   ROCK DOVE +
RUDDY V 6   WHITE-WINGED DOVE S, R, V, P ++
TV ++   MOURNING DOVE R S V P +
COOPER'S HAWK P 1   GROUND DOVE P 3
RED-TAIL 5   INCA DOVE P 1
BLACK HAWK P 1   GR. HORNED OWL P 2
SPARROW HAWK S 2 P 2   LESSER NIGHTHAWK 4 EN ROUTE
GAMBEL'S QUAIL P +   WHITE-THR. SWIFT P 8
SPOTTED SANDP. V 2   BLACK-CH. HUMMER R +

In 1973, as a teenager, I embarked on a "big year" to see how many bird species I could find in North America, traveling (mostly by hitchhiking) from Florida to Alaska and back. It was an extreme form of the list-chasing game, but even during this year I continued to keep field notes—mostly built around lists of the birds seen each day, but with some additional notes on bird numbers and behavior.

(continued on next page)

evening. Those field checklist cards from the Audubon chapter were expensive—a whole dime—so rather than using a new one every day, I wrote down my day's list on notebook paper, copying the sequence and names from the checklist.

My early attempts to beat my own list totals in this birding game provided my first experience in writing field notes. It seemed logical to write down, alongside the bird list, not only the date but the weather and the sites I'd visited. It seemed logical to make a note of how many individuals I had seen, at least for some species. And of course I would make some kind of note on the side if I had found a nest, or observed some interesting behavior. In a way, because of my listing games, I was reinventing a rudimentary form of the Grinnell method of keeping field notes, and establishing the rough pattern for the way I would keep notes in the years that followed.

Indeed, it has remained a constant for me that daily field notes will usually include a list of the species observed and at least some indication of numbers of individuals. It has become almost automatic; I nearly always make the list the central point in my notes. But there are exceptions. There are times when I am so focused on one or two species that I don't make even a mental note of what else is around; this is especially true if I'm sketching birds for field identification studies. In such cases, a species list would be spurious, and the sketches serve as my field notes for the day.

During my late teens I was playing the game of bird-listing for all it was worth, and I even spent twelve straight months hitchhiking around the North American continent to see if I could break the record for most species seen in a year.[1] After that, my interest in the idea of birding as a sport began to fade. I have friends who can tell me exactly how many bird species they've seen or heard on the African continent, in Colorado,

Here is a sample from a day in the field with some of my Arizona friends. And yes, I can't deny it, the notes are written in red ballpoint pen. No comment.

16 OCT 2004 - HIDALGO CO. - WESLACO:

BUTTERFLY CHECKLIST FOR THE
LOWER RIO GRANDE VALLEY OF TEXAS
(Cameron, Hidalgo & Starr Counties)

A-ABUNDANT	O-OCCASIONAL
C-COMMON	R-RARE
U-UNCOMMON	X-<5 RECORDS

Common name *Latin name*
Abundance codes for the LRGV only

Skippers—Family Hesperiidae
Spread-wing Skippers—Subfamily Pyrginae
__ Belus Skipper *Phocides belus* (X)
__ Guava Skipper *Phocides polybius* (O)
__ Mercurial Skipper *Proteides mercurius* (R)
__ Broken Silverdrop *Epargyreus exadeus* (R)
__ Hammock Skipper *Polygonus leo* (R)
__ Savigny's Skipper *Polygonus savigny* (X)
__ White-striped Longtail *Chioides albofasciatus* (U)
X Zilpa Longtail *Chioides zilpa* (O) 1/-
__ Gold-spotted Aguna *Aguna asander* (O)
__ Emerald Aguna *Aguna claxon* (R)
__ Tailed Aguna *Aguna metophis* (R)
__ Mottled Longtail *Typhedanus undulatus* (X)
__ Eight-spotted Longtail *Polythrix octomaculata* (R)
__ Mexican Longtail *Polythrix mexicanus* (R)
__ White-crescent Longtail *Codatractus alcaeus* (R)
X Long-tailed Skipper *Urbanus proteus* (U) 2/-
__ Bell's Longtail *Urbanus belli* (X)
__ Pronus Longtail *Urbanus pronus* (X)
__ Esmeralda Longtail *Urbanus esmeraldus* (X)
__ Dorantes Longtail *Urbanus dorantes* (U)
__ Teleus Longtail *Urbanus teleus* (O)
__ Tanna Longtail *Urbanus tanna* (R)
__ Plain Longtail *Urbanus simplicius* (R)
X Brown Longtail *Urbanus procne* (U) 3/4
__ White-tailed Longtail *Urbanus dorysus* (R)
__ Two-barred Flasher *Astraptes fulgerator* (O)

1000-1230 AT FRONTERA AUDUBON / 1245 - 1330 AT VALLEY NATURE CENTER
__ Small-spotted Flasher *Astraptes egregius* (R)
__ Frosted Flasher *Astraptes alardus* (R)
__ Hoffer's Flasher *Astraptes alector hopfferi* (R)
__ Yellow-tipped Flasher *Astraptes anaphus* (O)
__ Coyote Cloudywing *Achalarus toxeus* (U)
__ Jalapus Cloudywing *Thessia jalapus* (R)
__ Northern Cloudywing *Thorybes pylades* (R)
__ Potrillo Skipper *Cabares potrillo* (U)
__ Fritzgaertner's Flat *Celaenorrhinus fritzgaertneri* (R)
__ Stallings' Flat *Celaenorrhinus stallingsi* (O)
__ Falcate Skipper *Spathilepia clonius* (R)
__ Acacia Skipper *Cogia hippalus* (O)
__ Outis Skipper *Cogia outis* (R)
__ Mimosa Skipper *Cogia calchas* (C)
__ Starred Skipper *Arteurotia tractipennis* (X)
__ Purplish-black Skipper *Nisoniades rubescens* (R)
__ Glazed Pellicia *Pellicia arina* (O)
__ Morning Glory Pellicia *Pellicia dimidiata* (X)
__ Red-studded Skipper *Noctuana stator* (X)
__ Obscure Bolla *Bolla brennus* (X)
__ Mottled Bolla *Bolla cylius* (R)
__ Golden-headed Scallopwing *Staphylus ceos* (O)
X Mazans Scallopwing *Staphylus mazans* (C) 1/-
__ Hayhurst's Scallopwing *Staphylus hayhurstii* (X)
__ Variegated Skipper *Gorgythion begga* (R)
__ Blue-studded Skipper *Sostrata nordica* (X)
__ Hoary Skipper *Carrhenes canescens* (R)
__ Glassy-winged Skipper *Xenophanes tryxus* (R)
X Sickle-winged Skipper *Eantis tamenund* (A) 15/3
__ Hermit Skipper *Grais stigmaticus* (O)
X Brown-banded Skipper *Timochares ruptifasciata* (O) 1/1
__ Everlasting Skipper *Anastrus sempiternus* (R)
X White-patched Skipper *Chiomara georgina* (U) 1/-
__ False Duskywing *Gesta invisa* (O)
__ Horace's Duskywing *Erynnis horatius* (O)
X Mournful Duskywing *Erynnis tristis* (R) 2/-
__ Funereal Duskywing *Erynnis funeralis* (O)
__ Common Checkered-Skipper *Pyrgus communis* (R)

__ White Checkered-Skipper *Pyrgus albescens* (C)
__ Desert Checkered-Skipper *Pyrgus philetas* (U)
X Tropical Checkered-Skipper *Pyrgus oileus* (A) 25/20
__ Erichson's White-Skipper *Heliopyrgus domicella* (U)
__ Turk's-cap White-Skipper *Heliopetes macaira* (U)
X Laviana White-Skipper *Heliopetes laviana* (C) 1/3
__ Veined White-Skipper *Heliopetes arsalte* (X)
__ Common Streaky-Skipper *Celotes nessus* (O)
__ Common Sootywing *Pholisora catullus* (U)
__ Mexican Sootywing *Pholisora mejicanus* (R)
__ Saltbush Sootywing *Hesperopsis alpheus* (R)
Intermediate Skippers—Subfamily Heteropterinae
__ Dyar's Skipper *Piruna penaea* (R)
Grass Skippers—Subfamily Hesperiinae
__ Pecta Skipper *Synapte pecta* (R)
__ Salenus Skipper *Synapte salenus* (R)
__ Redundant Skipper *Corticea corticea* (R)
__ Pale-rayed Skipper *Vidius perigenes* (O)
__ Violet-patched Skipper *Monca crispinus* (R)
__ Swarthy Skipper *Nastra lherminier* (X)
__ Julia's Skipper *Nastra julia* (U)
X Fawn-spotted Skipper *Cymaenes trebius* (U) 2/2
X Clouded Skipper *Lerema accius* (A) 5/1
__ Liris Skipper *Lerema liris* (X)
__ Fantastic Skipper *Vettius fantasos* (X)
__ Green-backed Ruby-Eye *Perichares philetes* (R)
__ Osca Skipper *Rhinthon osca* (R)
__ Double-dotted Skipper *Decinea percosius* (O)
__ Hidden-ray Skipper *Conga chydaea* (R)
__ Least Skipper *Ancyloxypha numitor* (R)
__ Tropical Least Skipper *Ancyloxypha arene* (O)
__ Orange Skipperling *Copaeodes aurantiaca* (U)
X Southern Skipperling *Copaeodes minima* (U) 1/-
X Fiery Skipper *Hylephila phyleus* (A) 69/70
X Sachem *Atalopedes campestris* (A) 2/-
X Whirlabout *Polites vibex* (C) 2/-
X Southern Broken-Dash *Wallengrenia otho* (C) 5/2
__ Delaware Skipper *Anatrytone logan* (R)
__ Common Mellana *Quasimellana eulogius* (O)

__ Dun Skipper *Euphyes vestris* (O)
__ Nysa Roadside-Skipper *Amblyscirtes nysa* (U)
__ Dotted Roadside-Skipper *Amblyscirtes eos* (R)
__ Celia's Roadside-Skipper *Amblyscirtes celia* (C)
__ Eufala Skipper *Lerodea eufala* (C)
__ Olive-clouded Skipper *Lerodea arabus* (O)
__ Brazilian Skipper *Calpodes ethlius* (U)
__ Obscure Skipper *Panoquina panoquinoides* (U)
X Ocola Skipper *Panoquina ocola* (C) 8/6
X Purple-washed Skipper *Panoquina lucas* (U) -/1
__ Hecebolus Skipper *Panoquina hecebolus* (R)
__ Evans' Skipper *Panoquina evansi* (R)
__ Violet-banded Skipper *Nyctelius nyctelius* (O)
__ Yucca Giant-Skipper *Megathymus yuccae* (O)
__ Manfreda Giant-Skipper *Stallingsia maculosus* (X)
Swallowtails—Family Papilionidae
X Pipevine Swallowtail *Battus philenor* (C) 4/2
__ Polydamas Swallowtail *Battus polydamas* (O)
__ Mylotes Cattleheart *Parides eurimedes* (X)
__ Dark Kite-Swallowtail *Eurytides philolaus* (X)
__ Black Swallowtail *Papilio polyxenes* (U)
__ Three-tailed Swallowtail *Papilio pilumnus* (X)
__ Aberdus Swallowtail *Papilio garamas* (X)
__ Thoas Swallowtail *Papilio thoas* (X)
X Giant Swallowtail *Papilio cresphontes* (A) 6/3
__ Broad-banded Swallowtail *Papilio astyalus* (R)
__ Ornython Swallowtail *Papilio ornython* (R)
X Ruby-spotted Swallowtail *Papilio anchisiades* (R) 1/-
__ Pink-spotted Swallowtail *Papilio rogeri* (X)
Whites & Sulphurs—Family Pieridae
__ Costa-spotted Mimic-White *Enantia albania* (X)
__ Tropical White *Appias drusilla* (O)
__ Mountain White *Leptophobia aripa* (X)
__ Checkered White *Pontia protodice* (A)
__ Cabbage White *Pieris rapae* (R)
__ Great Southern White *Ascia monuste* (U)
__ Giant White *Ganyra josephina* (O)
__ Falcate Orangetip *Anthocharis midea* (O)
__ Orange Sulphur *Colias eurytheme* (O)

Field checklists for birds have long been popular, and with the growth of interest in butterfly-watching, we are now seeing more local butterfly checklists as well. The format of this one for southern Texas doesn't allow much room for numbers or notes, but I wrote in numbers after the names, with a slash separating those identified at two different sites in Weslaco, Hidalgo County. Courtesy of Mike Overton and Shawn Patterson.

or on the African continent and in Colorado this year. But I'm not keeping track. So in terms of personal experience, my bird-listing days are mostly far in the past.

That's only with regard to birds, however. I have launched major listing attempts with other groups of organisms several times in more recent years. When I moved to Arizona, an earlier curiosity about reptiles was rekindled by the diversity of lizards in the desert. I couldn't easily find a checklist of local species, so I compiled one by looking at the range maps in *A Field Guide to Western Reptiles and Amphibians*.[2] For a couple of years I was actively pursuing a herp list, going out of my way to find species of snakes and lizards and toads that I hadn't identified yet. In Arizona and later in Ohio, I went through periods of trying to run up big lists of butterflies and, despite the identification difficulties, moths. So far I've been too lazy to really buckle down and learn the damselflies (they're a lot harder than dragonflies), but eventually my friends will shame me into adding up my odonate list for Ohio, and then I'll be out there trying to figure out the spreadwings and sprites and dancers to add to the tally.

In all of these cases, it would be easy enough to see the downside, the focus on superficial aspects. But I see positives as well. The listing game has repeatedly provided me with a stimulating and effective framework for learning about the diversity and current classification within a group and for getting out to encounter some of these things in the field. I have no qualms about recommending this approach to anyone who is just getting started with a particular area of natural history: start a life list, start keeping day lists. By the time you come to your senses and stop, you will have learned a lot.

There are ways in which listing can also benefit the community, not just the individual, and contribute to total scientific knowledge. The activities of recreational bird-listers have added immeasurably to our detailed knowledge of bird distribution. At least in North America and Europe, if we look at range extensions, isolated populations, low-density patterns of migration or vagrancy, and so on, the overwhelming majority of discoveries in recent decades have been made not by professional ornithologists but by amateur birders.

I witnessed a perfect example of this in Arizona in the 1970s. A couple of birders who had moved into the state began promoting the idea of competing on county lists. Particularly in Maricopa County (centered on Phoenix), the birding community took the idea and ran with it. People were competing on their county all-time lists, county year lists, county day lists. Previously, Phoenix birders had been content to drive farther afield, to southeast Arizona to observe mountain birds, to the Colorado River to observe water birds. The pressure to find these things within the confines of their county led to an amazing number of discoveries. Extensions of known breeding ranges, previously unknown wintering populations, a host of new information about scarce migrants, all were turned up as by-products of the game. More than fifty new species were documented for the county in the space of nine years. The two avid young birders who started the craze went on to get doctorates in ornithology and are now professionals in the field, so their early game-playing didn't hurt their careers, and the avifauna of Maricopa County is now known much more thoroughly than it was before.

The idea of county bird-listing was hardly new in the 1970s. As long ago as the 1890s, Professor Lynds Jones, president of the Wilson Ornithological Society, was actively list-chasing in Lorain County, Ohio, and publishing his results in the society's journal, the *Wilson Bulletin*. Jones referred to his big day list attempts as "daily horizons," as distinguished from "censo-horizons," in which he tried for a big species list but also kept track of numbers of individuals. In 1899 he published a paper on "The Lorain County, Ohio, 1898 horizon," in which he reported the totals that he and W. L. Dawson had tallied: they had found 175 species for the year, out of a known county total of 221, and each of them had added to their own personal county lists.[3] Two issues later, he reported on a May 1899 effort during which he had recorded 112 species in a day in Lorain County,[4] a remarkable total for that era. Jones and Dawson both published many papers on bird distribution and were considered leading ornithologists of their day, and no one faulted their bird-listing games at the time.

This approach has been used with more than just birds. In the late 1960s, lepidopterist Keith Brown began systematically trying to maximize his daily species counts of butterflies at study sites in Brazil. The method proved so effective in revealing new information about local butterfly faunas that he published a paper about it.[5] Brown described how he got the idea "that the maximization of daily species lists of butterflies, a seemingly unscientific goal (though much employed in a sister area, ornithology), could give a large scientific fallout; and this has indeed proved to be the case." For example, he described how six weeks' effort in the Brazilian central plateau had turned up twenty-five species previously unknown for the area—but then he had adopted the "maximization" method, and in another six weeks, he had found nearly three hundred more new species.

Brown's surveys included collecting of specimens, of course (just as the surveys of ornithologists typically did in the late nineteenth century), and collecting will continue to be an important part of studies for many poorly known groups or regions. But the daily listing habit provides information that doesn't come from surveys based only on specimens. Collections are biased toward rarity. If a large regional butterfly collection contains very few Red Admirals for August, it doesn't necessarily mean the species is rare at that season; it may just mean that the species is so common that no collector pays attention to it by late summer. But a person who is trying to maximize their daily list is not going to ignore any common species. No matter how commonplace or boring, their find adds one more to the day's tally! This is one of the less-obvious values of keen listing: if a species is missing from the tally, it indicates that the person really didn't find the species on that day or in that place. It amounts to negative data rather than just a lack of data. This is particularly important for mobile or seasonal organisms, when one is trying to establish phenology or seasonal occurrence.

The listing games I've just mentioned (Jones in Ohio, Brown in Brazil) wind up contributing to general knowledge only if the people playing the games are also involved in more serious compilation of distributional data. Obviously this isn't always the case, and the checklists of many avid listers eventually get tossed out without having contributed to anything. But there are some programs wherein the sightings of birders can be used in more permanent databases.

One large-scale example began in the early 1980s, when volunteers were recruited to keep checklists of the bird species they had identified, in weekly increments and organized by county, within the state of Wisconsin. These birders filled out their lists on computer-readable cards and sent them in, first to the University of Wisconsin, later to the state's Department of Natural Resources. By 2007, over 94,000 of these reports had been turned in to the Wisconsin Checklist Project.[6] Even though the lists record nothing but presence or absence of a species, scientists have been able to get a rough index of timing of migration and changes in abundance of various birds by looking at the percentage of checklists on which a species occurs and tracking how this percentage changes over time.

There have been various other cooperative checklist projects involving birds and other organisms as well, but the most ambitious such project is eBird, a massive database being compiled by the Cornell Laboratory of Ornithology and the National Audubon Society. Made possible by vast computer capacity and the Internet, eBird encourages observers to report not only species but numbers of individuals, exact locations, exact amounts of time spent in the field, and additional notes on anything unusual. The amount of information pouring in is now phenomenal; a recent figure indicates that well over 50,000 complete checklists were submitted in an average month, and the numbers are still rising. Amazingly detailed maps of relative abundance for numerous species at all seasons can now be produced by eBird, and its capacity to do so will only increase as more birders take part in the program.

And more birders *will* take part, because the coordinators of eBird understand birder psychology. When the project was launched in 2002, scientists at Cornell and Audubon assumed that observers would be good citizens and report their information just because it was needed. Although response was lukewarm at first, the coordinators eventually hit on a way to capitalize on listing instincts of birders. If I log in today to report the birds I've seen during a day in the field, the welcome screen will tell me what my "life list" is. It isn't really my life list, of course; it just tallies the birds that I have reported to eBird. Farther down the page it will tell me how many species I have reported for this state, for this year, and so on. The incentives are clear to serious birders. Looking at these totals, I find myself thinking: "Well, I don't know what my life list would amount to, but I know it would be more than THAT!" I'm tempted to dig out and enter my field notes from trips to Alaska, Venezuela, Jamaica, and so on, just so those numbers will go up. Of course, if I do, that will also add to the total information in the eBird database. It's one more way that the game of listing can make a contribution, however small, to science. It's an odd combination, I admit, but it works.

A Reflection of the Truth

ROGER KITCHING

THE CITY OF HULL was, some would say still is, an unprepossessing place. I was born just outside that city in the tiny seaside town of Hornsea, to which my mother had fled to avoid the exigencies of the German blitzkrieg of 1944. Yet it was to the low-lying, fish-smelling, trawling port of Hull that we returned, and I spent my childhood and school years in that city. This was also where I became an obsessive naturalist.

The city was not, at first glance, an obvious place to become acquainted with wild nature—but first glances are often misleading. Back in the 1940s and 1950s this essentially flat city was crisscrossed with wide drainage ditches—the "drains," as we called them. Here we ventured with our fishing nets and jam jars to the discomfiture of sticklebacks and newts, tadpoles and frogs. Yet these modest waterways also introduced me to water scorpions, dytiscid beetles, and water spiders—even the occasional kingfisher—as well as a range of waterside plants on the boggy banks that bordered them. The city was also richly endowed with bomb sites—decaying rubble overgrown with weedy plants—but here were the swaths of rose-bay willow herb and the fat green larvae of the elephant hawk-moths which fed upon them; here were nesting finches and redstarts, jackdaws and thrushes; and here too was an ever-increasing range of shrubs and plants as nature reasserted itself once soil and space became available.

During my childhood years all of this gradually disappeared. The open drainage ditches were piped and covered, the bomb sites gradually cleared, and the city gentrified. Finally, with the collapse of the deep-sea

Notes and a drawing made of a Blue-banded Kingfisher (*Alcedo euryzona*) that I found, dead, beneath a window at the Danum Valley Field Station in Sabah, Borneo.

fishing industry, which fell victim first to Icelandic assertion of marine territoriality (the so-called Cod Wars of 1958 and 1972) and subsequently to changed rights of access associated with membership in the European Union, even the characteristic fishy smell of the city (especially when the winds were easterly) disappeared. Fortunately this progression was matched by one of my own as I discovered that most valuable of institutions—the local natural history society.

The Hull Scientific and Field Naturalists' Club was one of many such associations in the North of England that could trace their origins back to Victorian times and the "scientific" fervor engendered both by the industrial age and by the post-Darwinian revolution in natural history. By 1960 it had its share of professionals, but they remained, not surprisingly in this working-class town, a minority. They rubbed shoulders at each monthly meeting and weekend excursion with the true amateurs.

I recall in particular Denis Wade, a docker who spent his lunch hours (he would have called them dinner times) recording the butterflies and moths on the waste ground within the dock precinct, and Denis Pashby, as blunt a Yorkshireman as you could ever find, and also a docker, I think, who would catch the train or bus to Bridlington on the mere rumor of a Ross's Gull in the harbor or a Cetti's Warbler at Spurn Point. On one memorable occasion, Pashby traveled to the ploughlands of the Vale of Pickering to see the Common Crane, which had turned up seriously lost from its Russian migrations.

The "Field Nats" taught me many things. Most important, they let me know that I was not alone, that the inexplicable obsession with nature wherever and whenever it could be found was, if not common, at least not unique. I was also introduced for the first time to professional biologists, mostly high school teachers and university academics, who had taken their obsessions and made careers out of them. This sowed a seed that proved to be dramatically fruitful. I count among my first mentors the stalwarts of the Naturalists—Don Smith, Gwynneth Kemp, Frank DeBoer, Percy Gravett, Eva Crackles—who took a precocious but essentially untrained passion and began to give it direction and depth. Together with the biology master at my grammar school, Ken Fenton, these generous and knowledgeable people literally created my future.

Early in my career, I had discovered two technical aids to be essential if I was to make my way as a naturalist. The importance of these was reinforced by my early mentors—turning them into actions proved more challenging. The first was the amassing of collections. I was always an inveterate collector: first, but briefly, of birds' eggs, then of shells, fossils, wildflowers, fungus spore prints, feathers, and so on, culminating, when I was thirteen, with insects—butterflies, moths, beetles, bugs, and dragonflies. The current tendency to discourage youthful collecting, although well intended, is in my view wrong. Amateur collecting has almost never had any serious conservation impact—surely removing a small fraction of the caterpillars from a tree is not to be compared with removing the trees and all their inhabitants, yet we condemn the first as morally offensive and accept the latter as a perhaps unfortunate necessity. Insect collecting developed in me a wide range of technical skills, honed the observational abilities associated with identifying what had been collected, and of course gave structure and direction to the field excursions associated with collecting.

With collecting comes the need to record. It was impressed upon me from the very first that a specimen without a label was simply a (sometimes) pretty object. Without its associated data it was scientifically worthless. But there is only a certain amount you can squeeze onto that tiny rectangle of card on the pin beneath the butterfly. I realized at a young age that the circumstances of the object's capture, the weather, even the feelings of the captor and the experiences surrounding the event were important. I needed to record these not as a few telegraphic words on a label, but in narrative form—and this led me to the fine art of journal writing.

I had, of course, had the usual teenage brush with keeping a diary. The idea of writing down one's deepest feelings—it was never clear for what future purpose—is a commonplace. What is less common is sticking with it beyond those first few weeks after the arrival of the New Year's diary as part of the traditional Christmas booty. I did it assiduously for four years from the age of fourteen to seventeen. My deepest feelings, if this has any meaning at all for a teenage boy, were not what I recorded. Rereading my diaries through adult eyes reveals a stilted, mechanical narrative interspersed, but rarely, with opinion. And yet, as I read those diaries fifty years on, I can recall the events of particular days with a remarkable clarity. The diary provides the mental infrastructure that stimulates the mind to remember. This, it seems to me, is one of the great joys of keeping a diary and journal.

These childhood diaries contain accounts of many natural historical events and encounters, but they are not what I have come to call journals. Here there are no lists, no appended memorabilia and, perhaps most important, no drawings.

A childhood as an obsessive naturalist fortunately chimed well with the demands of the formal education system of the day. Back in the 1950s and 1960s, the English education system was both meritocratic and traditional. The abilities to read and retain, to write grammatical prose in at least two languages, and to regurgitate all of this at great speed in exams were much prized. All of these skills were equally essential for my development as a naturalist and, as this was clearly close to my heart, transferring them to the educational arena was easy. The constant creative writing associated with keeping a diary no doubt also helped. There

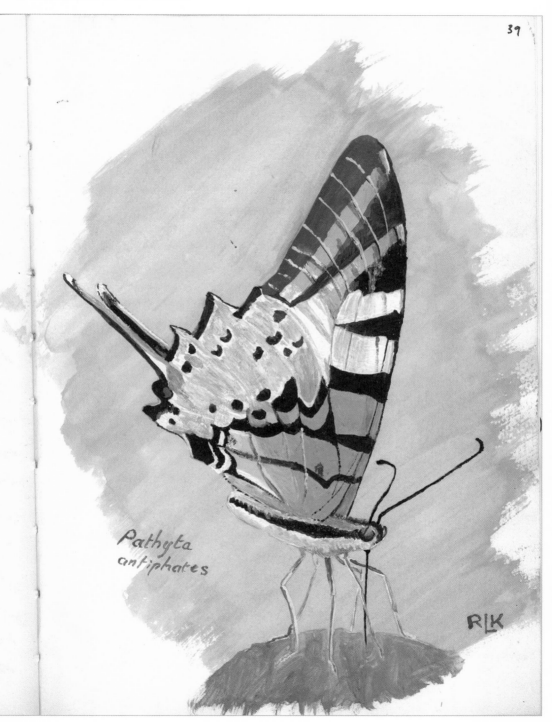

Pathyta
antiphates

R|K

A watercolor of *Pathysa antiphates*, a butterfly from Borneo also known as the "five-bar swordtail."

was a great deal of chicken-and-egg in this interaction between natural history and a formal education, but it all worked out rather well.

In those post–Second World War years, Britain had great need of a meritocracy to cope with the ever-increasing technical demands of the modern world. It did this through a competitive school system that often blighted the lives of the formally unsuccessful. However, it did create opportunities that would previously have been totally out of reach to those of my class background. These opportunities finally took me away from the council estates and foggy flatlands of Hull to Imperial College, London, to read for a degree in zoology and entomology.

This was a course of study that was essentially unchanged from when it was first designed by that great naturalist Thomas Henry Huxley in the late 1800s. Indeed, when refurbishing the preservative in some of our teaching specimens (a Port Jackson Shark, as I recollect), out fell a tiny identification label with the enigmatic initials *THH* attached. We spent two years working systematically through *every* phylum of the animal kingdom (with bits of botany and chemistry thrown in by adjacent departments). Then in our final year we repeated the dose by working through all twenty-nine Orders (as then recognized) of the class Insecta. This may seem both forbidding and unimaginative for a modern student, but for me, as a naturalist, it was perfect. All that inchoate information I had absorbed through catholic but undisciplined reading as a child was now given a logical framework. I now knew how the hemichordates related to the vertebrates, why fleas and flies are considered sister-groups, and why anatomical simplicity in parasites may represent a highly derived version of much more complex ancestors.

Imperial College was a great supporter of student expeditions, and the zoology students in particular (and there were only twelve of us in each class) were in great demand. We lent scientific credibility to excursions by the hearties from engineering or mining who wanted to head off for the biggest mountains, greatest glaciers, or just the most distant places they could find. This demand led me to spend long and instructive summers first in Ethiopia and second in the Arctic, where my first brushes with producing expedition-based journals took place.

One thing led to another and in 1966 I transferred to the University of Oxford to begin research for a D.Phil. Here I joined the Bureau of Animal Population (BAP) just before the retirement of its founding director, Charles Elton. Much has been written about the BAP and Charles Elton and though this is not the place to write extensively of either institution, some comment is essential. Charles Elton was the father of animal ecology in the English-speaking world. His 1927 book *Animal Ecology* is still essential reading for budding ecologists, and his later work *The Ecology of Invasions by Animals and Plants* remains the basis for many modern research programs.[1] Most important for me, though, was the emphasis on fieldwork—I just overlapped with the period when Oxford actually had a Department of Zoological Field Studies of which the Bureau was a part.

Elton was an avid note-taker, and my period at the Bureau coincided with the completion of his (or, actually, his secretary's) great task of typing up his notebooks, principally those associated with his work with the ecological survey of woodlands near Oxford (written up so elegantly in *The Pattern of Animal Communities*[2]). But that was not all. A lifetime's worth of notebooks included those dating back to his groundbreaking expeditions in Spitzbergen in the 1920s. Although Charles Elton was my research director during that magical period in Oxford, H. N. ("Mick") Southern was my actual thesis supervisor. Mick was a mammalogist, so his technical input to my essentially entomological thesis was limited, but he was also editor of the *Journal of Animal Ecology* and possessed an Oxford classics degree in addition to his zoological qualifications. So what he taught me was both the imperative to and the means of writing scientific prose—"if it's not published it's not done," as a later adviser put it. Mick showed me that the rather dry technical requirements of scientific writing did not necessarily mean that elegance, humor, and even wit need be excluded from the scientists' products.

These Oxford experiences and all my subsequent experience as a biologist have gradually converged into a three-layered process of documentation. First, there is the field notebook. This is where the actual numbers are recorded, together with passing observations relevant to the interpretation of these numbers. Even with the rise of laptop-based spreadsheets, I still try to ensure that all the data collected in the field make their appearance in a notebook as well as on that spreadsheet. Paper is still proving more durable than electronic data and ink more

A page from a field notebook made during a long-term research project in New Guinea on the food webs of aquatic animals known as *phytotelmata* that live in plant containers, such as tree hollows and bromeliad tanks. I use notebooks to record the quantitative aspects of fieldwork as well as observations relating to those records.

permanent than the simple polarization of electrons. For a start, ink does not readily change its state in the presence of a magnetic field, and the human optical recognition system (that is to say, reading) doesn't change in basic design every few years. Accordingly, the notebook remains an essential backup tool. It also works in all climates, without a source of electricity.

For me the journal is a parallel record to that of the notebook. Strictly speaking, journals are just daybooks for recording business transactions on a regular basis. But in a natural historical context, and certainly in my mind, they have become more associated with journeys—the common root is significant. Accordingly, when I began to make more and more frequent expeditions to far-flung corners of the globe, it was automatic that I attempt a daily account of events, thoughts, and observations.

My first attempt was a sad failure. During the 1965 Imperial College expedition to Ethiopia—a trip most influential in forming me as both a traveler and a scientist—my diary sadly peters out about halfway through the three-month expedition. I can remember what happened subsequently, but the detail that the day-by-day accounts now bring to mind is not matched by the episodic recollections that are all that remain without those aide-mémoire. As I age this disparity becomes more and more obvious.

Journals are parallel records to my notebooks that keep an account of my field-work in narrative form. This section is an excerpt describing work on container habitats in New Guinea.

27. i. 1988
A light breakfast, lab. chores (another Toxorhynchites & the first anthropogonid today), & then off to Baitata.—Again I was dropped there while the rest went diving—returning about four hours later.—an eventful four hours.—I collected 11 tree-holes which was very good. One rot-hole was at about chest height. It contained litter & water as usual.—I took my usual water sample by hand—and then noticed that the topmost leaves were moving slightly.—Assuming a frog I began to remove leaves from the surface—fortunately with long

forceps—whatever it was stirred & then 2-3 coils of a snake's body broke the surface, turning over in the way snakes have. I leapt back several feet. Then, reasoning that it would be a harmless water-snake I dived in with the forceps—and pulled out 3 feet of very angry elapid !!! — dropped, hastily, to the ground, I expected it to flee. But no, it reared up & struck not once but about six times in my general direction, mouth wide open & fangs visible. Extending the butterfly net to full length, I finally chivvied it out of the way. In due course I collected my sample & continued.—adrenalin-enriched if nothing else!

Collected butterflies also, although it was overcast & not the best collecting situation.—Got one or two nice lycaenids along the road while waiting for the car.

On the way back passed banks of red-flowering Clerodendron—frequently

Subsequent attempts were more disciplined, beginning with a 1966 expedition to Lapland. A long gap then ensued while I dealt with establishing a career, a family, and a scientific reputation, then in 1985 I participated in the Royal Entomological Society of London's centenary expedition to Sulawesi—"Project Wallace." I put real effort into the journal for that two-month period. After that, I forced myself to continue until writing those accounts at least every two or three days became habitual. I now have a shelfful of accounts ranging across activities on six continents, and they have become by far the most treasured part of my personal, professional archive.

Before leaving the general topic of journals, let me just add that my field journals were never written with direct publication in mind. In some cases, such journals are published because the writers are of such distinction and historical importance that making available every scrap of written material that survives may be justified. Even in these cases the general reader of Darwin, for example, will do far better to read the smoothed-over and selected material derived from the actual journals and published as *The Voyage of the* Beagle[3] than to plough through the slow-moving, often telegraphic day-by-day accounts now available.[4]

Last of the three strata, then, are the publications. Traditionally, in science, these are articles in academic journals leavened with chapters in books. To be successful, a young scientist need aspire to no more than these two forms of output together with their oral versions at interminable conferences and meetings of learned societies. Arguably, if these are the limits to your aspirations, then expedition-style journals may be superfluous, or, at best, a precious self-indulgence.

There came a time in my scientific development, however, when other forms of publication became important. The first of these is writing for popular, if specialized, magazines. Here the drive is to get across accounts of your own or someone else's research work. A perusal of articles in some of the better-known of these magazines is evidence of how hard many scientists find this task. The goal is to be clear on the science, but to leaven the accounts of results with anecdote and incidental detail. Here is where one's journals may prove immensely valuable. It may be scientifically interesting to demonstrate that food chains in tropical communities are generally longer than in their temperate counterparts (see below), but this is made far more interesting and approachable to a general reader if this deep truth is embedded in stories of discovering pitcher-plants on a floating bog in Sulawesi, or contending with the mud and steep terrain of a Bornean rain forest.

This becomes even more important, and the journals even more useful, when it comes to writing books. Curiously our present scientific administrations, at least in Australia, give little if any professional credit for the production of a book, and in some spheres, working scientists are actively discouraged from diverting any of their limited time into producing them. But in hindsight, this may be misguided. By way of example, the theory of island biogeography—the first area of community ecology to be given a firm theoretical base—was indeed first published in now largely forgotten articles in *Ecology* and *American Naturalist*, but the little book by Robert Macarthur and E. O. Wilson published in 1967 by Princeton University Press remains essential reading for all ecologists to this day.[5] There are many other examples.

In my own case, translating many years of work on invertebrate food webs into *Food Webs and Container Habitats* was a personal mission over a six-year period culminating in its publication by Cambridge University Press in 2000.[6] Here I delved deep into previously published papers, into notebooks containing unpublished data, and, most important, into a series of my journals. In these journals lay the incidental details by which a book can be differentiated from a set of scientific articles. Such incidental details can become ends in themselves, and the modern autobiographical travelogues that some ecologists are now producing are excellent cases in point. The fine book by Bill Laurence on his adventures as an American ecologist working in Australia is particularly pertinent.[7]

So having established, I hope, that journals, notebooks, and the published word are complimentary devices for recording the experiences and thoughts of naturalists, I provide an example of what my journals, and stories derived from them but intended for publication, actually look like.

AN INCIDENT AND ITS ACCOUNT

In this section I present a story derived from my travels and recorded in my journals—first the longer, polished account suitable (I trust) for wider publication followed by the actual text of the journal entry that was the source of the account. This is an example of how I use journal accounts as the core of subsequent writing activities.

ACCIDENT ON THE BELALONG

"Daniel has broken his leg!" These words, panted out by the wet and disheveled Campbell, were not what I wanted to hear. I particularly didn't want to hear them one day into a twelve-day field trip to the Ulu Temburong of Brunei in Borneo. I had seventeen other students to look after, many on their first trips overseas or to the tropics, and for all of them the first time deep into primary rain forest, reachable only by Iban longboat.

Nevertheless, after a moment's hopeful disbelief, a response was needed. At the time the Kuala Belalong Field Studies Centre had no emergency evacuation procedure. The field station's director was away for a couple of days and, if anyone was in charge at all, it was the cook. So simply throwing the emergency to constituted authority was not an option. The supposed victim was ten minutes away, a rough scramble to the top of a precipitous cascade that flowed through a series of pools and further waterfalls into the Belalong River. Campbell's shout roused everybody and there was a general movement toward the scene. "Stop," a voice of undeniable command rang out. It was not mine. Tim, my son, has been a member of the State Emergency Service in New South Wales for some years. He had trained in emergency rescue and had even appeared in the film *Lantana* in a staged rocky rescue. He had been included with this group of second-year undergraduates from Griffith University to give him the opportunity to visit normally inaccessible places in Borneo.

So at that point, Tim took charge. He sent all but four of the group to their waiting lunch, instructing them to come up to the accident site in thirty minutes, but not before. So Tim, Melinda, Campbell, and I grabbed rope, a first-aid kit, and the station stretcher and headed for the top of the Sungai Ikan, the precipitous creek where Daniel had fallen. When we got there we found him sitting on top of a huge moss-covered boulder, clutching one knee to his chest, the other leg stretched in front of him, and rocking back and forth in shock and pain. He immediately begged us for painkillers we didn't have.

After interviewing the victim, we learned that an hour after the safety briefing the whole group had received—a briefing that abjured them to move step by slow step in the steep, slippery, unfamiliar rain forest so as to avoid damage to both themselves and the rain forest—he and Campbell had decided to cross the head of the creek. At this point the water was funneled between two massive boulders through a one-meter gap. Daniel decided this was a short leap and with a young male's certainty of immortality, he jumped. The receiving boulder was covered in wet moss and he slipped, sliding down the face of the boulder into a deep pool about three meters below. Nothing but an undignified wetting would have followed but for one thing. A recent tree fall had laid a trunk across the creek at the base of the boulder. His left leg slid behind this, twisting him around and snapping tibia and fibula in a threefold compound fracture just below the knee. In one way he was lucky (although an appreciation of this did not emerge until much later). The bones broke forward, emerging briefly at the front of his lower leg. Had fate decreed they had broken backward, across a region rich in arteries and veins, then it is likely he would have simply bled to death. Thankfully, there was little bleeding.

At the scene Tim undertook the gruesome task of straightening out the leg, strapping the two legs together so one would act as a splint to the other, and getting Daniel tied to the stretcher while some of the anesthesia of shock remained. Then we faced the thorny issue of how to get him to the serious medical help he so obviously needed. Getting him downriver to the nearest longhouse and roadhead, Batang Duri, was the only option. But the normal way of getting from the creek to the landing stage at the Field Centre was over a ridge and then along a narrow and muddy track perched halfway up the steep riverbank. Access had been "improved" by a series of those single-log Iban bridges into which a few footholds have been hewn. These are not designed for heavy carrying, nor can they accommodate more than one person abreast. So we decided to take him down the rest of the Sungai Ikan to its confluence with the Belalong, where he could be loaded onto the longboat, a portion of the journey now being organized by the cook. This was a journey of about 200 meters, but it involved a series of short waterfalls, rapids, and deep pools.

By this time the rest of the group had arrived, together with about half a dozen of the Iban staff of the Centre. Tim organized them into two parallel lines, some perched precariously on boulders at the side of the waterfalls, others (the locals were good at this) maintaining virtually invisible footholds on the near-vertical rocks themselves. Yet others stood waist-, chest-, or even neck-deep in the pools at the base of the chutes. So we passed the laden stretcher from hand to hand, with those relinquishing hold scrambling around to join the lower end of the caterpillar. I have never before or since seen such an inexperienced group of people rise to the occasion so well. I remember Rohini, a delicate Indian student of perhaps 1.4 meters height in a pool with her mouth just above the waterline, taking her share of the weight with arms fully stretched, then scrambling out to do it again. I remember Amanda holding Daniel's hand, reassuring him and continually repeating pulse rates to Tim as we went—"52," pause for reassurance, "50," pause, "47," pause, "45," pause. This was the lowest point, and later Tim confided in me that he seriously doubted we were going to make it at this stage. But by then the procedure was established, the river was in sight, and rather forced, awful jokes began to flow. "48," "50," pause, "54," pause—some sort of crisis point had been passed.

Meeting the actual confluence of the Sungai and the larger river was perhaps the trickiest moment. The stretcher had to be maneuvered between the two final boulders while being swiveled such that it could fit lengthwise into the longboat. These Iban longboats are designed for going up and down the fast, shallow rivers of Borneo's interior. They are plank-built (these days, anyway) with squared-off prows and sterns. About eight meters long, they are just wide enough for two people to sit side by side, albeit in challenging intimacy. They are driven, usually with consummate skill, from the rear, where a powerful outboard is mounted. A second person stands on the prow with a long pole. His or her job is to fend off rocks, assist by poling, punt fashion, in shallow stretches, and to yell warnings back to the motorman whenever an unknown obstacle is spotted. The laden stretcher, lengthwise, just about filled the width of the boat. Tim, Campbell, and I climbed in and the rapid downriver journey began.

By now, Daniel had learned that by hooking the heel of his injured leg over the end of the stretcher and pulling he could take bone-upon-bone pressure off the break. He later pointed out that the boat trip, for him, was the most comfortable part of the whole incredible rescue—but "most comfortable" should probably be seen here as a comparative statement only. At Batang Duri, forewarned by radio from the Field Centre, a local ambulance was waiting for us, together with two paramedics. At this point I breathed my first sigh of relief and imagined the worst was over. The stretcher was lifted into the ambulance and Tim and Campbell returned upriver. The paramedics had a thing called an inflatable splint, which they wrapped around the damaged limb, causing Daniel considerable agony. It then turned out neither of them had ever used one before and didn't know what to do. I later figured out that there are two holes in an inflatable splint. With one you pump air in. The other, at the other end of the splint, is for quick release and needs to have a stopper in it during inflation. There was no stopper. Further, we discovered that the ambulance was made for the locals, not for six-foot, burly Australians, and the rear doors could not be closed properly without jarring Daniel's long-suffering limb.

So we set off on the eleven-kilometer ride to the local hospital in Bangar, the capital of Temburong Province. Provincial roads ending at Iban longhouses are not high priority for either construction or maintenance, and this road was worse than most of its kind. Every pothole, every ridge, every rock was, it seemed, hit squarely, and our patient was actually howling with pain by the time we got to the hospital. Once there, things did take a turn for the better, starting with a hefty shot of pethidine (the morphine drip Daniel was craving didn't actually eventuate until the next hospital), and he was attached to a saline drip. I gathered that, in 2003 at least, the little hospital in Bangar was basically concerned with maternity, child care, and the minor scrapes and bruises of everyday life in an agricultural area. The lesion of Daniel's break did get roughly cleaned and the leg was X-rayed, but further treatment was clearly beyond them. So a "heli" was called from the capital, Bandar Seri Bagawan, to evacuate the patient (and me) to the central hospital.

Sure enough, in about forty minutes a military helicopter turned up with its highly efficient Gurkha crew (accompanied for some reason by a young woman and child, who seemed to be along just for the Friday afternoon ride). The helicopter was an aging Iroquois, veteran of the Vietnam War, but it did the job. Two army paramedics, clearly used to this sort of thing, transferred Daniel to a new style of stretcher, loaded him in the main part of the helicopter, and shut me into a sort of cupboard just behind with a separate door—I never understood quite why. We then lifted off and ten minutes later landed at Bandar International Airport. I remember the amazing views of the log-cluttered Limbang river as we flew over the Malaysian salient which separates one part of the Sultanate of Brunei from the other. At Bandar a sleek, modern, longer ambulance awaited us, and we drove to the equally modern and efficient central hospital, where Daniel finally got his longed-for morphine drip. I left him then in very good hands and began my much more mundane journey back to the group in the field.

There were two sequels to this otherwise rather grim story. The first is that, once I had left Daniel at the hospital, I contacted Allastor Cox, the Australian High Commissioner in Brunei, to report the rather unfortunate condition of an Australian national. Now diplomats come in for all sorts of criticism, and my experience of them around the world has been mixed, but on this occasion all were simply terrific. Allastor immediately made a call on his mobile phone and the machinery of citizen rescue took over. They contacted Daniel's parents, comforted him, and ultimately arranged for his medical evacuation back to Australia. Without them our whole field course would have been in jeopardy, and I certainly wouldn't have been able to return to the field. Allastor's beautiful and gracious partner Susila visited Daniel and organized chocolate biscuits, vegemite, and other Australian staples to keep up his spirits while awaiting evacuation.

The second corollary came at the field station on the last night of our course. I had taken groups of research volunteers to this field station many times. I had also conducted hands-on fieldwork there myself. This particular group was the second group of Griffith undergraduates I had taken to Kuala Belalong. They were the successors to a number of earlier groups, mainly Earthwatch volunteers, which I had taken to the Kuala Belalong Field Centre. On all of these occasions, the Iban staff at the station had been endlessly helpful and efficient, but always reserved. I got the impression that they regarded us as amateurs in their forest. We were welcome visitors, especially as we paid for the privilege of being there, but we were to some degree effete, dilettante, not serious.

Throughout the ten days that followed the rescue, they had been, as usual, coolly efficient. On the last afternoon, though, one of them sidled up to me and said, "Kuni wants to give you a barbecue tonight." So I handed over the twenty dollars necessary to buy the required "chicken" and then let events unfold. The whole group of us trooped down to the dining room from our little houses in the forest at about six. Sure enough, chicken wings were sizzling over hot coals, but also a long table had been set out on the veranda overlooking the river. Here was a bottle of illicit gin, a stack of cans of tonic, a bowl of ice, lemons, and, most significant of all, a gallon jar of the famed *tuak*—rice wine. And, as we arrived, longboats carrying the rest of the longhouse residents turned up—with ghettoblaster, karaoke capability, and music videos. It didn't take long for the party to get going. And this time there was no separation of "us" and "them." Daniel joined us in the form of a full-size cardboard cutout manufactured by some of the more imaginative students—complete with (fortunately unrealistic) wooden leg. We ate and danced until about ten, then all sat in the river in the highest of spirits. Kuni and her friend Suraya taught us the Iban hornbill dance, the graceful sweeps and curves of the ritual no little enhanced by gin and tonics (and the *tuak*). The inevitable conga line followed.

So why was this? Not just a reward at the end of a tough field trip but, I firmly believe, a recognition that something real and serious had happened in the forest and the group had coped appropriately and seriously. It was the only time I felt we might have been considered a little "Ibanised"—I hope so anyway.

Black Baza

Notes and a drawing of the Black Baza (*Aviceda leuphotes*), a raptor from Southeast Asia, recorded here in my journal describing work in Cat Tien, Vietnam.

BRUNEI 2003 (pp. 12–18), Friday January 17th 2003

So back to the Station—a little diarising, answering questions etc, until the whistle went for lunch—we started down the walkway when C—— came tearing down the path "Come quick D——'s broken his leg!" So ensued one of the most gruelling, exciting, inspiring five hours of my life—written description may not be enough!

First Terry and others went chasing off with me in pursuit but, it was immediately obvious that the person who knew what he was doing, and took charge smoothly and with absolute authority was Tim K. He began gathering up bandages, first aid kits, rope and the stretcher, ordered everyone to stay put for twenty minutes and then come after us.

I, meanwhile, walked across the boardwalk to the "Ladies' Bathing Pool" down the steep wooden stairs to the creek bed, then upstream—lots of rocks, two steep chutes with knotted ropes down them—to the very top of the waterfall where we found D——, clutching his left thigh with his lower leg dangling—he had slipped about two metres down slippery rock, caught his leg between rock and log, his weight carried him forward and, voila, a three point break—two in the tibia and one in the fibula and, as we later discovered, some knee damage. He was in great and voluble pain.

Tim simply told us what to do—with Ripley (recently retired from the Queensland State Emergency Service) as excellent back-up—sort of strange to see ones son with natural and totally authentic command—not, logically, surprising but known of, hereto, as in a different world. But this rescue was our rescue and totally here and now. We got D——'s legs straightened out and tied together—the good leg acting as a natural splint. Then he was moved onto the stretcher—with pillow brought up from KBFSC—a blanket and rope to tie him to the stretcher. Meanwhile Terry and I rigged extra knotted ropes down the two chutes. We dared give D—— only two Panadeine forts—indeed we have nothing stronger.

Then everyone else arrived—Tom, Tim, Cameron, Leanne, Amanda, Robin, Annie, Amy, Nicola, Erin, Wendy, Karen, Jen, David, Melinda, Sally and Lisa (Kimberley actually slept through it all)—plus Sujipto, Salleh, Ina, Surayah, Ramla and all the other staff. Directed by Tim we formed two double lines—continually renewing caterpillar-fashion—and, so, passed the stretcher down the 200 metres of the roughest imaginable terrain—the chutes were particularly challenging where the sure-footed Iban were vital—regular pulse checks—as low as 45 at one point improving to 55 when we finally got him to the boat—it took us two hours from encounter to laying the stretcher in the longboat, of pain for D—— and mighty, coordinated, beautiful effort on everyone else's part. The last rock squeeze was particularly bad with already soaked people chest-deep in water.

Once in the longboat, with a few dry clothes and our papers (organised instantly by Mel and Sally), Tim, C—— and I, with D—— and the boat person headed down rive[r]—D——'s pulse improved and steadied. At Batang Duri the little field ambulance was waiting with totally useless paramedics. The 18km or so from Batang Duri to Bangar were the worst of all according to D—— every bump in the road causing bone to move against bone—twice the ambulance stopped to try to adjust the situation—an inflatable splint they didn't know how to use, an incompetent nurse and an ambulance basically too short for the purpose. About 60 minutes it took us from long boat to Bangar hospital.

Once at Bangar things took a turn for the better, Tim and C—— headed back to KBFSC—I stayed with D——. A great doctor/senior nurse (I'm not sure which) gave D. a pethidine shot and began cleaning up the wound—not without some pain to the lad, as they felt, poked and prodded—he, occasionally, yelling abuse in English and Malay! They x-rayed him (thus we found out about the breaks) and stabilised him—put him on a saline drip—and called in the 'heli' as they described it. In no time at all (well maybe 45 minutes) the aged military Iroquois came into view and via their best ambulance we were loaded in and took the 20 minute flight to Bandar—two much more competent paramedics, three crew— and a woman and child who had come along for the ride. We landed at Bandar airport—another ambulance awaited us—thence, to the general hospital emergency and accident department. Here, finally, D—— got his morphine—that he had been yelling for 5h earlier! The orthopaedic surgeon was in theatre but the very efficient young male and female doctors exuded supreme confidence—he is to be cleaned up and disinfected tonight and decisions on the required operation on his leg made tomorrow. By now the patient was feeling decidedly chipper and the endless talking machine aspect was kicking back in.

So I took my leave.

LEGACY OR JUNK?

This example I hope illustrates my point that journals are memory prompts and perhaps capture exquisite (and not so exquisite) moments of experience. To publish these accounts demands much more incidental detail drawn from elsewhere in the journal, or from all the diverse sources—books, papers, Internet—at our disposal as writers. But it is the journal accounts that allow me to browse through my memories and select which incidents can provide a core around which a publishable account can be constructed.

So, finally, what is the value of these journals in the broader scheme of things? Perhaps each journalist will have a different answer. For me they are quintessentially personal accounts—scattered through my journals are personal snippets that are certainly not intended for any eyes but my own—during my lifetime at least (I doubt I shall be concerned one way or another after that). And yet, when the urge to write public accounts of some of the expeditions comes upon me, here is the vital memory extender. Here is a way of bringing to the forefront of my mind events, even incidental detail, which allows a dry, rather telegraphic account of events to be expanded into something that can have a wider appeal.

Finally, these are a legacy. One of the recent pastimes of my wife and me has been to research and reconstruct our respective family trees. We have succeeded to the point where we know the names of several hundred direct and collateral relatives going back seven or eight generations in some cases. We know the dates of their births, marriages, and deaths. Occasionally we know where they lived and what their occupations were. Yet we do not know these people. I yearn to know what they thought about life, where they went for their holidays (did they have holidays?), what they talked about, what they read (could they read?), what were their hopes and aspirations. If only they had left journals!

Perhaps that row of handwritten accounts, with their sometimes wobbly drawings and their bird lists, will be read by my grandchildren (currently too young to read) and give them some idea of what made the old man tick. I hope so!

Linking Researchers across Generations

ANNA K. BEHRENSMEYER

AS A PALEONTOLOGIST, I spend countless sun-baked hours exploring rocky landscapes searching for time capsules. What I discover is often just an unrecognizable fragment, but sometimes I find a breathtaking skull or a bonebed that immediately opens a window through time, showing what life was like millions of years in the past. The intense planning and gritty toil that go into this work are validated by the knowledge I gain from these specimens and the rocks that encase them. Some of this new understanding is distilled into scientific papers, which is gratifying to me and may be informative and even exciting to others. These publications, however, document only a fraction of the work that takes place on those long, hot days of exploration and discovery. While in the field, there is a continuous flow of information that consists of daily observations, insights, and data, and this has presented a problem for me over the course of my career. How can I capture these data and thoughts in notes, maps, and images so that they will be of value both to me and to future generations of scientists?

An important first step in solving this problem was my realization that in recording fieldwork I was creating my own time capsules. Studying paleobiology gives me a fundamental appreciation of the transmission of information across time—whether it involves fossils of extinct organisms or written passages conveying ideas or descriptions—and also an understanding that how well these capsules hold up over time makes all the difference. A career of fieldwork is minuscule compared with the expanses of time that I study. Yet, my observations provide a pathway back into the record of past life, a time machine where I interface with, and try to understand, small-scale outcomes of large-scale processes. A good field journal entry can take me back to the date and time of a major discovery or explain the circumstances of a particularly fruitful (or frustrating) day of field surveying, even after the passing of many years and even decades. A poorly kept set of field notes, on the other hand, is only a dim reminder of observations that cannot be recalled or recovered from days long past.

The lasting relevance, or "half-life," of a set of field notes depends both on the quality of the project—the scope, goals, and basic information—and on the way the observations and ideas are actually recorded. Charles Darwin's field notes have held their value over time because they are neat and meticulous and effectively transmit the essence of his revolutionary observations and thinking across centuries. Journals kept by other early explorers became best sellers because they conveyed a personalized sense of wonder and discovery along with massive amounts of information, much of which still has scientific value. There was a strong tradition of observation and record-keeping underlying these early documents, but that tradition has changed for more recent generations of field scientists. The myriad tools of the digital age that provide quick ways to capture words, images, and data have added to the perception that handwritten field notebooks are passé. As someone who routinely encounters objects that can speak to us over millions of years, I may have a bias toward things that have stood the test of time. That said, it is clear that there is still much to recommend preserving records and information in traditional paper field notes.

In spite of the expanding virtual world, the fundamentals of good fieldwork have not changed much since Darwin. Google Earth can take you to a field site anywhere on the planet, but there still is no substitute for actually being there, walking over outcrops and through time. Technology such as computers, GPS, and the Internet can similarly make information so abundant and easy to record in vast quantities that they become a substitute for sitting down and writing out clear and precise records. This scattered digital information breaks down over time, however, without the cohesion provided by a single, well-kept field notebook.

What excites me about fieldwork is that it always holds the potential for incredible (real—not virtual) discoveries, inspiring me to walk up another windswept hill in order to solve a geological puzzle or to hunt for buried treasure in the form of new fossils. It would be easy to assume that once I have snapped a digital photo and taken a GPS reading, these discoveries are adequately documented, and any remaining puzzles can be solved when I return to my lab. But experience has taught me there is no substitute for taking the time to question, puzzle, explore, and document observations and insights while in the field. Even when there is little to show for a day's effort, my fieldwork provides me with time to think in beautiful outdoor settings and to be inspired by present and past worlds. Being out together in the natural world also engenders life-long friendships and a community of colleagues who share the challenges and fun of doing fieldwork. Field notes provide direct connections to these thoughts, experiences, people, and the resulting scientific data, and if done well they will inform future scientists about our research in a way that is easily understandable and contributes to the scientific understanding of our planet.

The most effective way for me to achieve this goal is to spend time in the field with my eyes and ears open, and with a pen and notebook ready to record what I find. Over the course of my career, I have developed a habitual field note protocol in which a paper notebook is used both to record information and to integrate records made on standardized datasheets, in computer files, and in photographs.

If I had adopted this system from the very start, it would have saved me huge amounts of time searching for information as I tried to link details that were clear in the field but faded quickly after returning home. Although I had some early training and good role models, much of what I learned about note-keeping resulted from trial and error, evolving over several decades to where it is today. This system became extremely important because any scientist, and any granting agency for that matter, wants to get as much as possible out of a project, and carefully conceived plans for taking and preserving field notes provide a crucial link between the initial goals and ultimate contributions—enhancing both over the short and long term. Researchers who seek grant funds for a field project pay a lot of attention to the scientific scope and goals in their grant proposals but usually less on the protocols for field data collection and stabilizing data relevance for the long haul. Today, successful grant proposals usually strike a balance of exciting science, sound methodology,

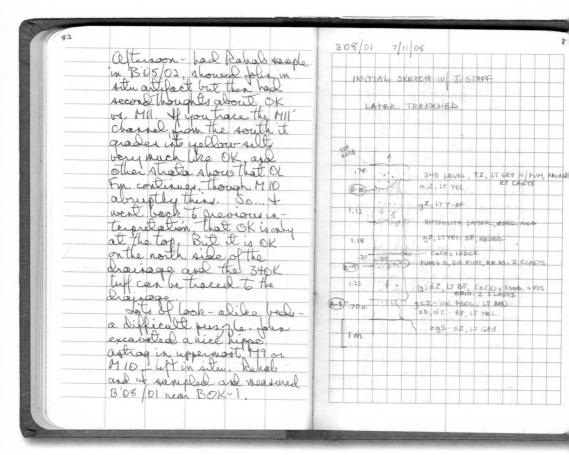

This example shows field notes taken while sampling phytoliths at Olorgesailie, Kenya, with Rahab Kinyanjui of the National Museums of Kenya. These notes were recorded in a durable orange-covered bound notebook. The right page has a diagram of the strata showing where the samples were taken.

and careful planning for documentation and archiving. Most funding organizations require databases to be permanently archived on accessible websites, and a well-honed system of field notes, including scanned images of the notes themselves, can contribute to the value of these databases.

Overall, I'm glad to have invested all those hours writing in my field books and am happy with the notes I have made over the years—notes that I refer to daily as I continue to write up the research. However, if I could go back to the beginning of my career and give myself advice about this process so that it could speak effectively across time, I would emphasize five basic rules.

RECORD YOUR WORK AS NOTES
TO YOUR FUTURE SELF AND COLLEAGUES

Field notes, and laboratory notes too for that matter, should be written with posterity in mind. Documenting almost any conceivable field experience may be important as a record of an observation in nature, a new insight about the way the world works, or even just a step in the evolution of your own thinking and education. You may have no idea about the future significance of these experiences when they are happening, and it is far better to assume that they will be of interest to someone in the future (especially you) than to think that they will not be so.

I first confronted the importance of keeping good field notes at Indiana University's geological field camp in southwestern Montana. My experience as a student there, and in the following summer as a teaching assistant, taught me how to document complicated geological information as we learned about mapping strata on aerial photographs and recognizing structural discontinuities. We were required to take observations at sequentially numbered "stations" as we spent whole days roaming over wide areas, finding our way using the aerial photographs, knocking off pieces of outcrops with our rock hammers to identify the strata, and gradually piecing together the history of the land under our feet.

Looking at strata in a deep pit, dug below an archaeological site at Olorgesailie, Kenya. Note camera and orange field book. Photograph by John Yellen (2008).

Part of our grade was based on the thoroughness and accuracy, as well as the organization and legibility, of our field notes. I still have the notebooks, diagrams, and marked-up aerial photographs from those times, and they take me back instantly to the days when I was exploring a new career as well as working out the geological history of a small piece of Montana.

One of the most important lessons from that early instruction at the IU field camp is to write notes so that someone fifty years from now (or more) will understand and be able to use the factual information you collected, perhaps for purposes quite different from the original reasons you undertook your fieldwork. Fieldwork today includes GPS technology, digital satellite imagery, field-adapted laptops, and other high-tech equipment, but it is still critical to use your notebook as a single synthetic record of what you are doing each day. Think of the field book—which can be pulled out at any moment (rain or shine), never runs out of batteries, and can be dropped without breaking—as the essential binding that holds together all of the information you collect.

As one example of why this is critical, I always record GPS coordinates in the notebook as a backup in case the digital files are lost or corrupted. That way, I can get back to the place where I recorded important information or found something unusual. Anyone who is interested can consult my notes in the future and will know what I did on that day, and also how to find the exact spot where I was using only one source—the field notebook.

In addition to factual observations, it's fine to record your initial scientific interpretations and personal impressions, but remember to clearly separate facts from interpretations so these are not confusing to a future reader. It is also fine to record thoughts and experiences that will help you remember each day or field area and enhance the interest of your notes. Remember that objectivity is important for communicating to others through your written words. Avoid subjective or strongly opinionated statements about logistical difficulties, negative experiences, and the shortcomings and failings of other people. The future credibility of your field notes will be compromised if they are sprinkled with inappropriate language, gossip, and other evidence that you were distracted from the tasks at hand. Fieldwork can include challenges in dealing with teammates, weather problems, and unexpected setbacks.

Detailed "microstratigraphic" diagrams such as this one from Olorgesailie, Kenya, provide the important baseline information about artifact and fossil-bearing layers of Pleistocene sediment. The circled numbers provide a reference for each bed, and the circled "R" numbers indicate where Rahab Kinyanjui took samples for fossil phytoliths. I do a lot of these on separate pages on a clipboard, as well as in my field notebooks.

95

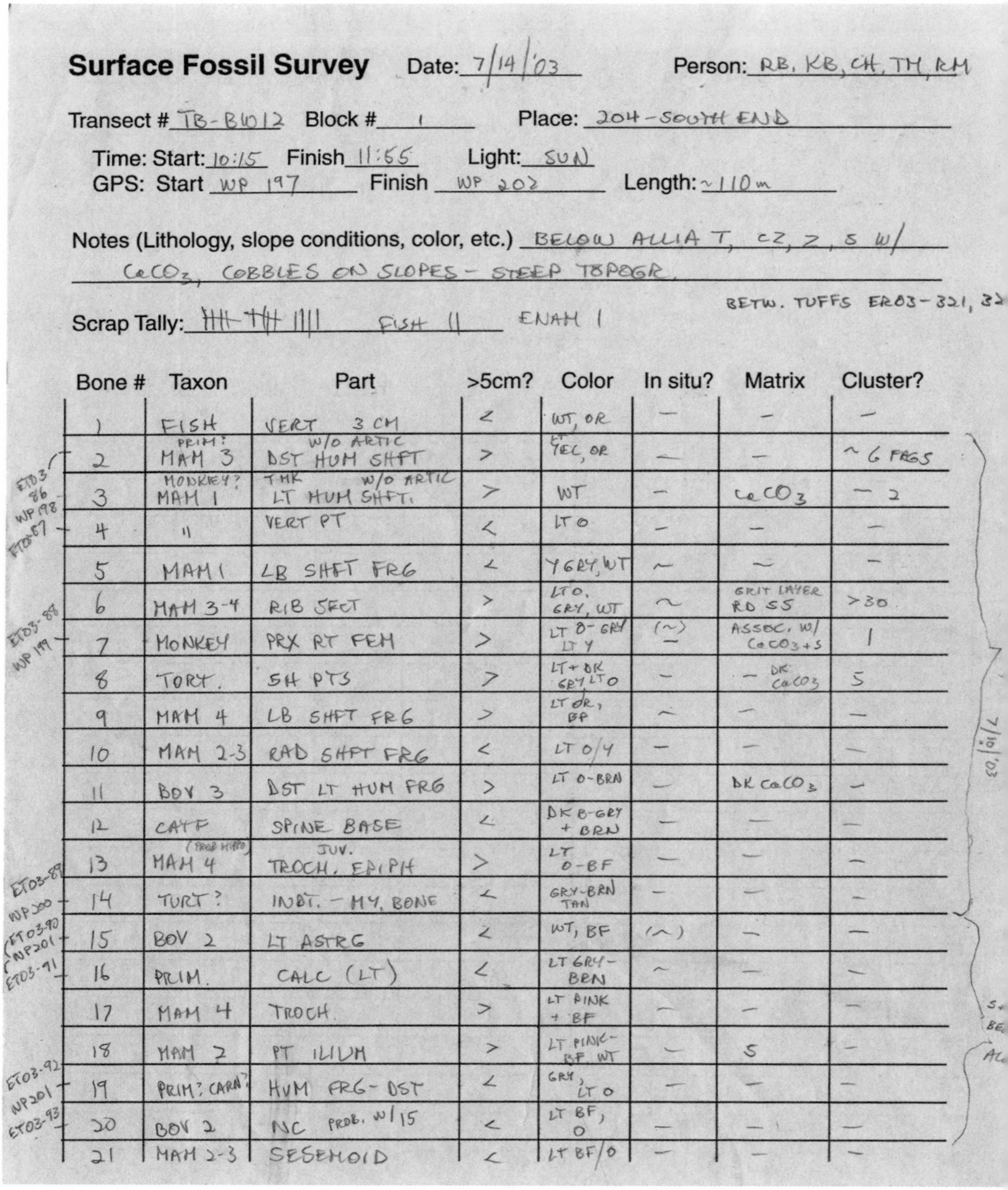

Surface Fossil Survey Date: 7/14/03 Person: RB, KB, CH, TM, RM

Transect # TB-BW12 Block # 1 Place: 204-SOUTH END

Time: Start: 10:15 Finish 11:55 Light: SUN

GPS: Start WP 197 Finish WP 202 Length: ~110 m

Notes (Lithology, slope conditions, color, etc.) BELOW ALLIA T, CZ, Z, S W/ CaCO3, COBBLES ON SLOPES - STEEP TOPOGR. BETW. TUFFS ER03-321, 32

Scrap Tally: ~~||||~~ ~~||||~~ |||| FISH || ENAM |

Bone #	Taxon	Part	>5cm?	Color	In situ?	Matrix	Cluster?
1	FISH	VERT 3 CM	<	WT, OR	—	—	—
2	PRIM? MAM 3	DST HUM SHFT W/O ARTIC	>	LT YEL, OR	—	—	~6 FRGS
3	MONKEY? MAM 1	THK LT HUM SHFT. W/O ARTIC	>	WT	—	CaCO3	— 2
4	"	VERT PT	<	LT O	—	—	—
5	MAM 1	LB SHFT FRG	<	Y GRY, WT	~	~	—
6	MAM 3-4	RIB SECT	>	LT O, GRY, WT	~	GRIT LAYER RD SS	>30
7	MONKEY	PRX RT FEM	>	LT O-GRY LT Y	(~) —	ASSOC. W/ CaCO3+S	1
8	TORT.	SH PTS	>	LT + DK GRY, LT O	—	— DK CaCO3	5
9	MAM 4	LB SHFT FRG	>	LT OR, BF	~	—	—
10	MAM 2-3	RAD SHFT FRG	<	LT O/Y	—	—	—
11	BOV 3	DST LT HUM FRG	>	LT O-BRN	—	DK CaCO3	—
12	CATF	SPINE BASE	<	DK B-GRY + BRN	—	—	—
13	MAM 4 (PROB HIPPO)	JUV. TROCH. EPIPH	>	LT O-BF	—	—	—
14	TURT?	INDT. — MY. BONE	<	GRY-BRN TAN	—	—	—
15	BOV 2	LT ASTRG	<	WT, BF	(~)	—	—
16	PRIM.	CALC (LT)	<	LT GRY-BRN	~	—	—
17	MAM 4	TROCH.	>	LT PINK + BF	—	—	—
18	MAM 2	PT ILIUM	>	LT PINK-BF, WT	—	S	—
19	PRIM? CARN?	HUM FRG-DST	<	GRY, LT O	—	—	—
20	BOV 2	NC PROB. W/15	<	LT BF, O	—	—	—
21	MAM 2-3	SESEMOID	<	LT BF/O	—	—	—

ET03-86 WP198 ET03-87; ET03-88 WP199; ET03-89 WP200; ET03-90 WP201; ET03-91; ET03-92 WP201; ET03-93

Example of a standardized field data collection form used to record all the fossil bones encountered along a transect that follows outcrop surfaces along particular strata. Informally I refer to these as "bonewalks." Some of the fossils were collected (notes on the far left) but most were not. Having this type of information provides an unbiased record of the kinds of animals that make up the fossil assemblage and is an important counterpart to traditional fossil collecting that focuses on the best-preserved identifiable specimens.

It is worth mentioning logistical difficulties, but it is not good to indulge in daily personal commentary when what will matter in the long run is a carefully kept, objective record of your field science.

Some notes of a personal nature add color to the field records. I have frequently mentioned when it was a "really good day," "long, tiring day in the rain," "bad luck—two flat tires," and similar comments. These often stir up memories and help me recall important details about collections or the field site. On the other hand, I once learned a lesson involving a field logbook that was available for everyone in camp and was supposed to be a group record of what was going on with daily research activities and discoveries. Anyone who wished could make entries, and it was assumed by the camp director that these entries would be responsible and objective. When some

A typical notebook page detailing the thoughts and events of a day doing fieldwork at Olorgesailie, Kenya, with a personal note near the end of the page about the joy of being alone with the rocks.

amusing but somewhat derisive comments appeared in the logbook, the director removed it and the daily records came to an end. One moral of this story is that it is better to keep individual rather than group notes, and what you write always should reflect a high standard of professionalism to readers now and in the future. At the end of each day in the field, it's good to question whether what you have recorded will be understandable many years in the future, and make additions or corrections until this way of writing becomes second nature.

ESTABLISH A CLEAR AND CONSISTENT
NOTEBOOK FORMAT AND PROCESS

When you are just starting out and are not responsible for a scientific project, taking notes may not seem particularly important. My first real paleontological field experience was in the Wind River Basin of central Wyoming when I was a beginning graduate student, and the expedition leader did the note-taking while the crew did the finding and digging. The badlands we searched were Paleocene in age—about 60 million years old—and the eroding strata contained fossils of strange animals from the beginning of the Age of Mammals. Previously, in addition to geology field camp, I had gone on many outdoor adventures. But weeks of continuous tent camping with a few other people on the banks of a dry, cottonwood-lined streambed was a new kind of experience. We would go out every day prospecting for fossils, mapping, and recording geological information, and would dig up anthills where the ants had carried small pebbles, including fossil teeth. These little insects helped us by concentrating tiny fossils as "armor" on their anthills, but digging into their mounds resulted in some painful bites. The fossil deposits contained a puzzle that spurred me on to what became a lifelong interest, taphonomy, which is the study of how organic remains become fossils. Buried together in the ancient sediments we explored were fossils from plants and animals that lived in very different environments. We unearthed charcoal from trees, bones from ground-dwelling mammals, and shark teeth. How did all of these remains become mixed and interred in the same place? Figuring this out has led to years of fascinating research, but at the time it did not occur to me that keeping my own notes would be a good thing to do. I have often wished that I could

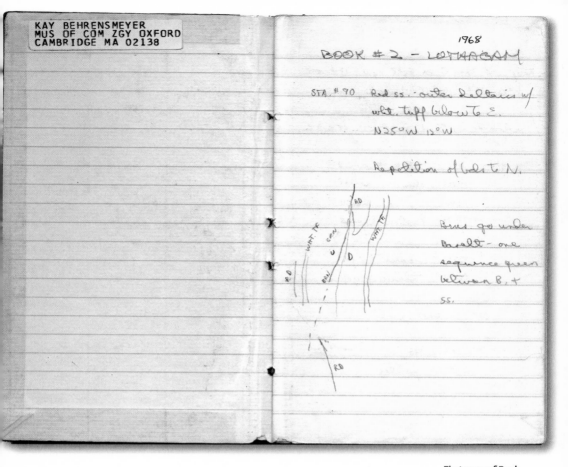

1968

BOOK #2 — LOTHAGAM

STA. #90 Red ss. · outer deltaics w/
wht. tuff below 6 =.
N25°W 15°W

Repetition of beds to N.

Beds go under
Basalt – one
sequence green
between B. +
SS.

return to my ideas from that time—another hard lesson on the road to becoming a committed note taker.

Three months after leaving Wyoming, I traveled to northern Kenya where I joined a colleague and two Kenyan assistants on an expedition to a remote, hot, and dry area near Lake Turkana. This expedition was an adventure of the first order. We were in the field for five weeks, far from any town, and we had to haul all of our water in a trailer. I learned to get by on a ration of one cup of washing-up water per day because we had to save all we could for drinking and cooking. Pulling the trailer back to the nearest town to resupply involved a day's work digging wheels out of the dry riverbeds where our vehicles regularly got stuck in the hot sand. Our assistants took care of most of the camp chores, so I was freed from helping with cooking, supply runs, and car maintenance. The two scientists were able to focus entirely on the fieldwork, and working out the geology of the area was my responsibility. The site was rich in fossils,

First page of Book #2, one of my earliest (1968) Kenyan field notebooks from Lothagam Hill, using a locally bought school primer book. The sketch shows geological relationships in plan view; a fault marked by U (up) on one side and D (down) on the other slants across the drawing. It would have been good to put an approximate scale on the sketch; this was one of many lessons learned by experience about taking field notes.

and we collected many well-preserved remains of Mio-Pliocene verte-brates, including extinct pigs, antelope, carnivores, hippopotamuses, and crocodiles. I mapped the strata and figured out the geological age of these unique remains. At the time, this site (Lothagam Hill) was the only known source of fossils for this age in East Africa, between 7 and 4 million years ago. It was of great interest to paleoanthropologists because a puzzling jaw fragment and some teeth were found there, representing an ancient hominid that possibly was on the lineage leading to modern humans.

At Lothagam, I kept my own records and many of these are still informative, but they are inconsistent and leave much to the imagination. I wish that I had chosen a clear and consistent format as well as more durable notebooks and recorded information every day. The notebooks I used were purchased in Nairobi and were made for schoolchildren— they were never intended for field use or long-term survival and are now falling apart. These days, I make sure I choose one with a brightly colored cover so I can find it easily, and one that is the right size for my belt pouch, so it is always handy. I always use a pencil for sketching, but I write the daily journal entries with a black ballpoint pen so that these cannot be easily altered. If I make a correction, I lightly cross out the original and write in the new one. This leaves a trail in case I actually got it right the first time!

Including your name and contact information on the first page and cover, along with the project name, place (continent, country), and the year will help you and other readers in years to come. On the first pages of my notebooks I describe the overall goals of my fieldwork and list all the people in the field team, visitors, local contacts, nearby landmarks, and towns. It is also useful to sketch a draft calendar for scheduling activities and meeting goals for your time in the field. When I use symbols or codes to cut down on the writing, I am always sure to provide a clear key to these in the notebook. If I'm not likely to remember what these represent years hence, it is even less likely that someone else will figure out their secret meaning.

I take some notes every day, even if nothing much happened. Missing days will make future readers wonder whether I left out something important or just forgot to take notes on a period of "business as usual."

I have gotten into the habit of assessing the day's results and writing them down, and if things are slow, I revisit research goals and record thoughts about progress and plans for the coming days and weeks. For each day's entry, I always include the date, place, main activities or events, weather conditions, and other people involved. The day, month, and year is the most important link between that particular point in time and other people's records, separate data sheets that I filled out myself, photos, and most important, collected specimens. I never forget to include the date, even putting it on each page in case separate pages are scanned or photocopied.

Although my daily activities are always recorded (where I went, what I did, whom I did it with), documenting collecting strategies and protocols receives special attention. In the moment, these may seem like common knowledge for the field team, so sometimes no one bothers to write them out. Several years later, few if any team members will remember whether we agreed to collect teeth and jaws of pigs and only pigs on that day, or whether we were collecting all the fossil bones we saw (even fish and turtle scraps). The result is that we cannot always be sure whether there was a collecting bias or not, and relative abundance tallies for interesting animals (such as giant pigs) versus other species are useless without this information. This

Title page of one of my latest notebooks, with important information about the contents of the book and the GPS datum, as well as a full address in case the notebook needs to be returned to me.

issue of collecting bias has plagued paleontologists for generations. Of course paleontologists usually go for the "good specimens" or target the species of greatest interest to the project (such as early humans—finds that help to fund future research!). This is fine, but without a written record of such collecting protocols, future generations of scientists will not know how to treat the apparent overabundance or absence of certain types of organisms in the collections. In Kenya, there were a few years in which fossil pig teeth were targeted for collecting because researchers discovered that their morphology changed steadily over time, making them good indicators of the age of the deposits. If you plot the Kenya National Museum catalog records of pigs versus other types of mammals, an odd abundance peak appears for pigs in the 1970s. This is a mystery unless there is someone around who remembers the directive to collect all well-preserved pig teeth. Since the people bearing this kind of knowledge can disappear or forget, a written record of such "obvious" or "assumed" protocols is one of the best things a field researcher can do to make data and collections useful over time.

I find that numbering the pages of my notebooks makes it easier to cross-reference earlier work. If I run out of notebook pages before the year is up, I glue in additional pages to avoid splitting years across two notebooks that may become separated. If I must use two notebooks in a given year, I clearly indicate the presence of the second at the end of the first.

If you use a GPS, be sure to record the geographic datum that is specified on the navigation page (very different depending on where you are on the planet). Not knowing this in the future may seriously compromise your GPS data, because different datums can generate up to hundreds of meters of difference in position coordinates and distances. I also record instrument settings for surveying and other types of field equipment, such as magnetic declination for a compass.

When I try to locate and piece together fragments of related information from my many notebooks, rock and fossil collections, and thousands of field photos, I can see how it would have been good to forge even stronger links among these with consistent note-taking from the early days of my career. On the brighter side, however, I'm now discovering how to reconstruct these linkages by inventorying the information in the notebooks and organizing this in multirelational spreadsheet formats that allow more cross-referencing than would ever be possible with the notebooks alone.

DON'T LOSE YOUR FIELD RECORDS!

There is almost nothing worse than losing the only copy of a field notebook with important, hard-won data. I have never had this misfortune, but I know a number of researchers who have, and in some cases it resulted in tragic loss of information and nearly derailed their projects and degree programs.

Aside from preventing the loss of your notebook, there are other ways that you can avoid losing information if you think beyond your primary responsibilities on a research team. My first notebooks from the Lothagam project in Kenya were very tightly focused on the geological goals, following the procedures I had learned at the Montana field camp. It turned out that I should have included more about other aspects of the work and day-to-day experiences, especially the fossil finds. The field catalog of fossil specimens we collected was kept by a colleague, who did all the daily entries. Later, when that catalog went missing, other researchers asked if I could help. Had I kept more information on the fossils themselves—especially linking the numbers on the specimens with places where they were found—that redundancy would have allowed us to reconstruct this vital information. The catalog remains missing to this day, and there are many unresolved questions about important specimens from that expedition. This taught me that more comprehensive note-keeping, along with creating multiple records and archive copies, are more than worth the extra time they take. It is better to err on the side of too much seemingly redundant information than too little.

Before leaving any field site, I cross-reference what I am doing in the field with the work of other researchers, photocopy, scan, or photograph the pages of my notebooks, and leave one copy with the host museum or a trusted colleague. I always carry my original field book in my hand luggage. Similarly, if I take a field notebook out of my office to the field area, an archive copy is left at home.

This photograph was taken on an expedition to the beautiful Kayenta Formation of central Arizona, where I documented the geology of rich fossil deposits in order to reconstruct the paleoecology of the earliest North American mammals, which lived approximately 200 million years ago. A tiny black human in the middle left area of the photo shows the scale of this vast area.

PACK A CAMERA, CREATE A VISUAL RECORD

I mourn the extinction, though perhaps temporary, of the Polaroid camera, as it has played a vital role in my fieldwork over the years. No matter how many words you write to describe a fossil locality, you can't beat an actual photo, taken on the spot, annotated in pen, and pasted into your notebook.

In my paleontological work in Pakistan, where the team searched a huge area of eroded landscape looking for fossils and recorded the sedimentary layers shed by the rising Himalayas, it was essential to be able to identify each of hundreds of localities as well as the places where I measured the strata. My field notebooks are filled with instant photos showing where we worked. In addition to the notebook entries, we used a "locality card," which was filled out for each separate fossil locality. A photo—black and white, which doesn't fade—was attached to each card, annotated in the field to show where fossils were found (including the direction of the photo, date, time of day, and locality number). The project left a large box full of the locality cards in Pakistan and kept a duplicate set at our home institution. The card plus marked locations on topographic maps or aerial photographs allow someone who has never previously visited a locality to find it and see where the fossils were collected. Since we now have over 1,600 localities, this systematic protocol is absolutely

A page from my 2007 field notebook showing the use of annotated Polaroid photographs to understand the paleoecology of the Kayenta Formation of central Arizona.

essential. Even with today's GPS capability, it is important to have photographs that show where fossils were collected, and these images provide a backup in case of errors in the GPS readings.

We can adapt to the loss of the Polaroid camera by taking along a small printer and digital camera (hoping that it functions in rough field conditions) in order to have printed photos in the field. Digital field notebooks with touch screens also may help to solve the on-site documentation problem, though such equipment is typically more delicate and fussy than the old instamatic. A computer screen can be difficult to read in bright sunlight, however, and one always has to worry about charged batteries.

Using cameras and annotated images, whatever the technology, will allow you to include in your notebooks photos of fossils in situ, field localities, special geographic features, and other visual information that will help people in the future understand the spatial relationships of field specimens or other objects relative to strata, excavation grids, and vegetation types. Images are also an important tool for enhancing descriptions of plants, animals, and habitats, and I usually tape or glue them into the field book. It is also a good idea to take high-resolution digital images for later, more formal presentations of information from the field-annotated photos.

In the future, instant photo prints likely will require field workers to carry a printer and digital camera, paper and batteries, and to protect all of this from heat, dust, sun, and rain. This will probably lead many to decide to do the printing in camp, annotating from memory unless they go back to the site. However, I still believe that there is no substitute for a photograph you actually mark in "real time" in the field as the best way to preserve a lasting, accurate record for yourself, or for someone who has never seen the site or object in question. I am also not convinced that digital printouts will archive well, so I hold out hope for a new generation of instant cameras.

LEARNING THROUGH SKETCHES AND DIAGRAMS

Photographs are great, but drawing what you see is a more powerful way to learn about spatial patterns and relationships. Thought diagrams, such as flow charts, food webs, or time lines, are valuable ways to conceptualize research questions whatever your area of specialty. Much of the documentation in fossil-finding involves recording the rock strata associated with the specimens, so it is essential to make "cross-section" diagrams of the layers of strata to show the fossil-producing level or levels and, if possible, their lateral patterns along the outcrops.

In biology or ecology, sketches can be equally valuable in showing geographic features associated with a sampling site, structure of vegetation in a habitat, the positions of traps or other sampling devices, and so on. Even if you are not an expert at drawing, you can make sketches that are much more informative than words would be. But even if your talents are limited, always include a scale (preferably metric), even if it is a crude estimate, and always indicate north, or up versus down, or the direction to some known place, so that a person fifty years from now

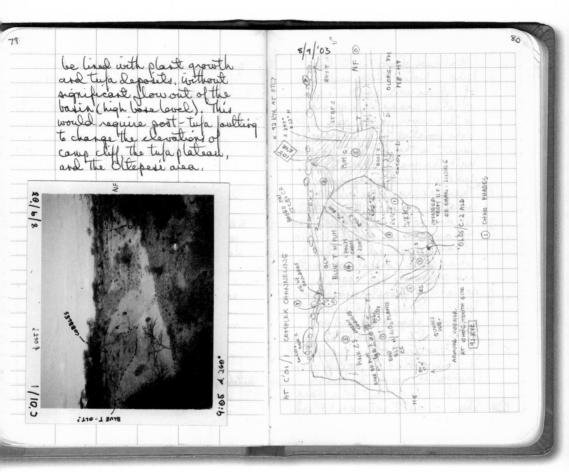

can reorient and understand the size of your diagram relative to the real thing. It is also important to label your diagrams so that you can remember what they mean. Sketches can be redrawn later or used in combination with photographs or Google Earth to create more presentable versions of maps or perspective images.

CONCLUDING THOUGHTS

All of us have been encouraged, sometime in our lives, to write down thoughts and feelings in personal journals or diaries. A field notebook is a special kind of journal that adopts the rigorous standards

A typical notebook entry from Olorgesailie, Kenya, in 2003 diagramming the complex geological relationships of Pleistocene volcanic channel deposits (righthand page). The Polaroid shows the same area, but only through careful observing and drawing can one figure out the details of such outcrops.

of science while also providing a unique record of our personal experiences as scientists. I haven't counted lately, but my career so far must have generated at least fifty notebooks from four continents and many different countries and field sites. Like the rock strata I dig through in the field, I still mine these notebooks for information, peeling them back page by page, and it is good to know that they will become part of the Smithsonian's permanent archives when I retire. I cannot imagine banks of computer files being as accessible or having the power to take me, or other unknown future readers, back to those great days in the field, to the facts, colleagues, excitement, and insights that are still alive in the handwritten pages of these special journals.

SOLAR 2/19 BODEGA ⑭ ⑬ cf. CASA

6

The Spoken and the Unspoken

KAREN L. KRAMER

HUMANS ARE COMPLICATED. As an anthropologist who works with humans living in traditional societies, I seek to understand aspects of the human experience that relate to how we evolved as organisms, how societies are constructed, and how they function. Living in traditional societies in remote areas of Mexico, South America, and Madagascar, I have forged relationships with people who have ways of life very different from my own, with whom I live very closely, and who are also the subject of my research. My field notes are contained in maps, data sheets, notebooks, and journals that approach this experience from different angles and attempt to address both the qualitative and quantitative aspects of human behavior. Ethnographic studies are distinct from ethological research in other species because we can speak with our subjects and ask them questions. This has tremendous value, but much of what humans do is not spoken, and we also observe, count, and measure. Over the years, I have developed a multifaceted approach to field notes that allows me to record facts, ideas, and observations about my study populations. From these records, research questions emerged that I never expected when I was making them. But before describing my own process and experiences, I will briefly situate how my approach has been molded by fundamental challenges in the study of anthropology.

ANTHROPOLOGY IN CONTEXT

The field of anthropology was built on the study of hunter-gatherers, horticulturalists, and pastoralists who lived in small-scale societies throughout the New and Old World. When explorers, naturalists, and cartographers began casting about the world on the heels of military and religious conquest several hundred years ago, they wrote about the people they encountered in these traditional communities as well as the geography, plants, and animals of the surrounding habitats. These ethnographic records and field notes have become much more rigorous and focused over the last several centuries, and this parallels a similar transformation in biological documentation. Early ethnographic accounts were anecdotal and often embedded in the narratives of explorers and naturalists. Fascination with exotic cultures spurred florid descriptions often intended to either horrify or delight readers. As anthropology distinguished itself as a specific discipline in the latter part of the nineteenth century, descriptions became more probing and detailed, and highlighted the historic and cultural uniqueness of different societies. Evolutionary thinking shifted focus toward the relationships between environments and adaptive variation. This connection between environment and human biological and cultural variation remained largely descriptive through the mid-twentieth century. But as anthropologists working in different parts of the world began comparing ethnographic notes, it became clear that more exacting field methods were needed to make meaningful cross-cultural comparisons. Beginning in the 1960s, observational methods borrowed from biology—especially primatology—substantially impacted the collection of ethnographic information.

Concurrent with this trend toward more systematic methods, the populations that anthropologists traditionally studied were rapidly disappearing as they transitioned to more acculturated ways of life. These groups became immersed in cash economies through wage labor, experienced permanent settlement as well as the introduction of food production, market foods, craft production, vaccination, health care, and contraception. Many anthropologists began to work with these populations and in urban settings with new interests and research questions. For anthropologists asking evolutionary questions, however, traditional societies continue to provide important insights into the range of human biological and behavioral variation not represented in modern, well-fed, low-fertility, low-mortality populations.

Anthropologists and biologists share many field and analytic methods as well as many of the logistic rigors of field research. But there are also differences that impact the nature of their documentation and field notes. In human research, because both the researcher and our subjects use speech to communicate, anthropologists can depend not only on observation, but also on interview methods to collect data. While verbal communication and conversation about past events with subjects has many advantages, it also introduces novel elements of bias and deception, resulting in additional record-keeping challenges.

QUANTITATIVE AND QUALITATIVE
APPROACHES TO DATA COLLECTION

The numerous ways in which anthropologists today collect data and keep field notes can generally be distinguished as being qualitative or quantitative in character. Qualitative observations provide many descriptive details, personal impressions, background information, and provocative anecdotes that inspire research directions and are important to breathe life into the people we study. Quantitative data collection involves systematic and repetitive observations on the same set of variables. Compiled, quantitative observations allow a researcher to consider not only the differences between things, but also how much they differ. These scale-level differences form the basis of comparative analyses. Although skepticism periodically emerges over the perfunctory nature of quantitative methods to explain human culture, if we seek to make comparative statements, at the most fundamental level we first have to be able to compare different kinds of apples and how they relate to different kinds of oranges.

My field methodology is guided by an interest in human life history and demography. Because my questions are comparative and quantitative in nature, my research depends on quantitative data. However, if I become too focused on systematic data collection, I can easily lose sight of the very people whose lives I seek to understand. Neither qualitative nor quantitative approaches tell the whole story. Instead, it is important to develop recording strategies that retrieve information relevant to specific research questions. To balance these two points of view, my field notes include a variety of formats, including hand-drawn maps, data sheets, and several types of journals.

GETTING STARTED

I work with three very different traditional societies—the Maya, a group of agriculturalists from Mexico; the Pumé, hunter-gatherers living on the *llanos* of Venezuela; and the Tanala, horticulturalists from highland Madagascar. My research interests are focused on behavioral and biological factors that influence demographic processes. The purpose of the Maya fieldwork, my first long-term ethnographic project, was to collect time allocation and demographic data. I was interested in understanding cooperative breeding and the high surviving fertility that characterizes human life history. Addressing this question involved documenting Maya mothers' reproductive histories and how mothers and children spend their time. However, the first step in my fieldwork was unrelated to my research question.

As a stranger arriving out of the blue, it would not have been a successful research strategy to immediately inquire about children's births and deaths and ask to shadow villagers throughout their day to record what they were doing. While the Maya are remarkably hospitable and may well have granted such a request, we would have all been stiff, and this would have set an uncomfortable precedent for a year's stay. Accurately capturing how people spend their time is contingent not only on systematic data collection, but also on participants moving in a relaxed and normal manner through their daily activities. Just as primatologists habituate their subjects to their presence, anthropologists first must develop rapport and trust with the communities in which they live.

One process I found that helped me become integrated into their community was to map the village. Like people everywhere, the Maya are intrigued by a bird's-eye view of their world. Constructing a detailed map of village households took several months, but permitted my field partner and me to move about the village and get to know villagers. The spatial organization of a village is an apt reflection of a society's social and kin organization. Through the process of mapping the village, I could determine who lived in the same house, who lived next door to whom, and what households worked the same gardens. This allowed me to

Map of a Maya village in the Yucatán, Mexico. Map construction helps in understanding the spatial organization of villages and the relationships among inhabitants, and also provides a way to develop relationships with villagers.

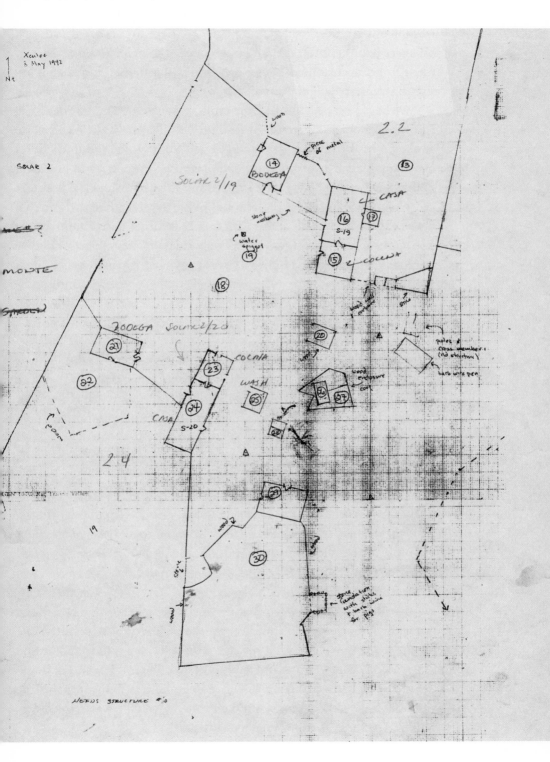

learn about kin ties, food sharing, and marriage and descent patterns. Villagers became interested and often offered to help, and this allowed them to get accustomed to our presence without our immediately intruding on their daily routines.

We began mapping during the rainy season, which in the Yucatán is stunningly hot and oppressively humid. The villagers watched us as we sweltered in the midday heat and slogged through the mud, intent on our mapping endeavor. Our fascination with instruments and precision under these conditions greatly amused the villagers, who have the good sense to schedule work at cooler times of the day. I learned early on that it was essential to laugh at our own foibles in order to live comfortably in circumstances where we were not only studying our subjects, but they were also studying us. Humor, often at my own expense, was a great conduit to developing friendships and becoming part of the community.

ASKING AND WATCHING WHAT PEOPLE DO

Many animal studies are concerned with how individuals allocate their time and energy to various somatic and reproductive pursuits. Because time can be systematically and repetitively recorded, it is a useful unit of measure to compare differences in usage across age groups, sexes, populations, and species. Anthropologists use both observational methods and interview methods to estimate time budgets. Each has its strengths and shortcomings.

Early in my Maya research I conducted household interviews. As with the mapping, it was a way to get to know people's names and ages, build the village census, and ask a few questions about how people spent their time so I could refine my time allocation methods before I started collecting behavioral observations. One of the questions that I asked was "How much time do you spend in the fields?" Several weeks after starting the interviews, I noticed that the response on the women's forms was always the same: no time. Yet every morning I saw women leave for the fields outside town. Later, the time allocation study would find that women spent a considerable proportion of their day working in the fields. If I had relied only on asking women how they spent their time and not observed how they spent their time, it would have revealed very different pictures about women's work effort.

Interview and recall data depend on participants accurately remembering their previous activities and on subjects answering the question that the interviewer thinks he or she is asking—a problem exacerbated when negotiating differences in language, cultural norms, and individual perceptions. For example, late one afternoon toward the end of my fieldwork, while I was recording behavioral observations, a mother turned to me—nursing a baby in one arm and tending the fire with the other hand—and said, "I've finished working for the day, you can go home if you want to." She was patiently fanning a fire to heat water that she would use to bathe her three youngest children, and she still had to prepare the evening meal of beans and tortillas for her family and settle her six children for the night. Yet from her point of view, the workday was over—a stark contrast to a conversation that might take place at the end of the day with an American working mother. I said that I'd like to stay a little longer and asked her if she didn't consider bathing and feeding her children work. She looked at me, puzzled by the question and the difference in our perceptions.

What is considered work may differ widely across cultures and among individuals. Recall methods, either through interview or self-reported in diaries, can be problematic in having to account not only for memory error, but also for individual variation in what participants report. For instance, if you ask children how much time they spent in school last week, will they include the time spent walking to school, the time spent doing homework, or the time spent at lunch or recess? Would each child answer the question in the same way?

One way to bridge the gulf between what people do and what they think they do is to directly observe their activities. Initially developed to record primate behavior, behavioral observation methods document the actions of the subject while in his or her company, rather than reconstructing them through interview or recall. Scan samples and focal follows are two commonly used behavioral observation methods. During a scan sample, randomly selected individuals are located at specified time intervals, usually every ten to fifteen minutes, and the observer instantaneously records what the participant is doing. After repetitive observations, an accurate estimate can be made about the proportion of time an individual spends in various activities—how much time they allocate to fieldwork, domestic work, child care, leisure, socializing, and the like.

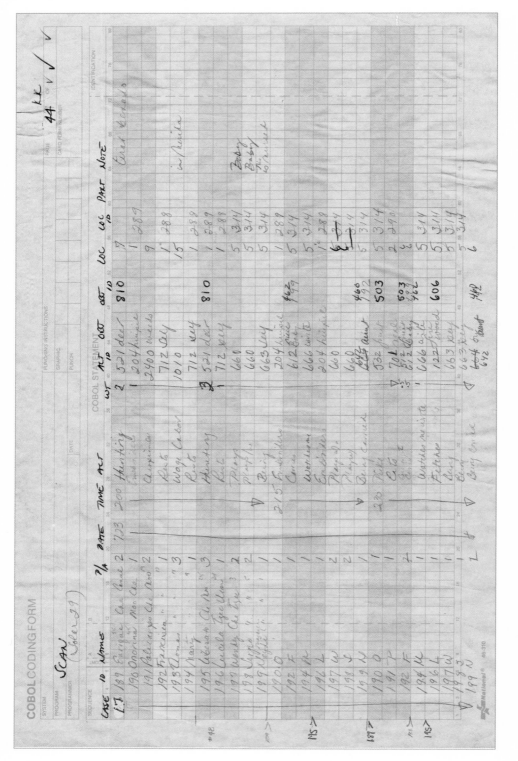

A "scan sample" data sheet from a Maya village is used to record what villagers were doing at set intervals of time (usually every 10–15 minutes).

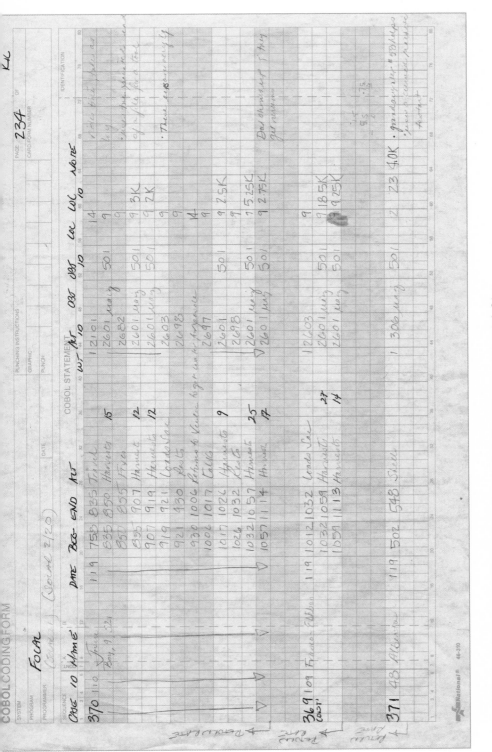

"Focal follow" data sheets contain continuous observations of a single Maya villager over a set period of time.

Focal follows complement scan samples by documenting the continuous sequence of an individual's activities. During a focal follow, each subject is observed over a period of several hours with each change in activity recorded with a start and stop time. A variety of information can be retrieved from focal follows such as return rates, the duration and periodicity of activities, sharing, cooperation among individuals, and food consumption.

Although behavioral observation methods more accurately reflect how individuals spend their time than do interviews or recall, they do not dispense with researcher bias and perception. For example, in most traditional societies children help care for their younger siblings. However, it is often the case that a child minding his younger sibling does so out of the corner of his eye while playing with other children. Is this play or child care? How it is systematically coded directly affects the extent to which children in a particular society are characterized as important allo-caretakers. I was keenly aware of this problem with the Maya, since mothers have on average seven or eight children, and young children are often in charge of their even younger siblings. Since I did not know what I might want to do with the play data in the future and waffled on how to classify this very common activity, my solution was to retain my indecision in my coding system.

I organized behavioral codes to contain several levels of information. As in this example, if a child is outside playing with friends while minding her two-year-old sister, the activity was coded as 675: the 600 signifies a noneconomic activity, the 70 that it is playing, and the 5 that it is playing while in charge of a child. All activities were coded in this way. A nested classificatory hierarchy preserves both detail for future research and flexibility to lump or disaggregate activities for analyses. This method of nesting information carries over into many kinds of coding and classificatory schemes.

Behavioral observation is an ideal method to collect a large sample of accurate data on multiple individuals. The downside is that it demands much more time from the researcher's perspective than interviewing subjects. While behavioral observation does not eliminate bias, it does have the advantage of minimizing the number of participant filters for which a researcher has to account. That said, the household interview did reveal an interesting point about how the Maya perceive the sexual division of labor. By cultural norm, Maya women do not think of themselves as field workers. This was a vantage point I would not have appreciated if I had collected data only through interview or only through behavioral observation.

FIELD NOTES AND COMPUTERS

Many new recording tools and technologies are now available to field researchers. Although field sites often do not have electricity, lightweight, portable, and inexpensive solar panels now make using a gamut of battery-powered equipment feasible. Each researcher has his or her personal preference on how to record field notes. For a number of reasons I favor paper forms and a pencil as the best compromise between enforcing consistency and maintaining flexibility.

Coding sheets prompt the researcher to systematically record all variables. The paper format also allows me to correct mistakes easily and jot down sundry narrative observations and notes. Although we have solar-powered computers and handheld data recorders at our research sites, I find it cumbersome and time-consuming to scroll through databases and menus while making adjustments and corrections. This distraction pulls my focus away from the situation, and deadens my awareness of subtle social interactions and cues to what people are doing. Although we take it for granted, our technological gadgetry is a curiosity to people for whom it is a novelty. Digital equipment is a sure way to interfere with the normal flow of activities. Any piece of equipment invites a steady stream of onlookers, drawing attention unnecessarily to the process of data recording itself. Writing is much less obtrusive.

Entering coding forms into a computer database at the end of each day, however, is invaluable. It catches recording problems and coding inconsistencies early on, and maintains a daily check for mistakes. Amending problems is much more difficult if errors are discovered months later when the circumstances of the day are blurred with other days, and can easily result in lost data. In situations where trips to town to use a photocopier are extremely rare, creating a backup also provides peace of mind.

NARRATIVE FIELD NOTES

In addition to the daily systematic data collection, I kept three types of journals to record when certain events happened, the machinations of my methodological decisions, and narratives of everyday life. While I did not anticipate their utility at the time, in the long run it was these narratives that often became critical to putting together many of the arguments I later would make with the data.

This calendar notebook documents Maya village activities and events chronologically.

In one notebook I kept track of village events—a record of the agricultural cycle, religious holidays, political meetings, when school was and was not in session, when the gas-powered well was broken, when merchants came into the village, and when medical teams visited. At the time, I recorded these events for their descriptive value and my own diversion. They since have become an indispensable guide to special circumstances that influence time allocation decisions.

I kept another notebook for coding definitions, explanations of why and when I coded something one way and not another, and why I made certain field sampling and methodological decisions. At the time, it may seem that these details will never be forgotten, but invariably they are. What seemed like a morass of trivia turned out to be key in reconstructing biases, verifying inconsistencies in data collection, and making the appropriate adjustments.

A third set of journals I kept for myself. Describing everyday life of the Maya world around me was a valuable way to appreciate what was an exquisite human experience, and also to maintain normalcy under circumstances that at times seemed far removed from my cultural frame of reference. I have been asked many times about what I miss when living so remotely. I rarely give the expected answer—movies, a shower, sleeping in a bed. I relish the simplicity of field life. However, I was not prepared for how much I would miss conversation in my own language. Writing in the evening helped me fill that void.

One of my early journal entries reads:

> Today while I was in town buying supplies, I found a book, the first Maya ethnography written by Bishop Landa in the 1500s. In it there is a map showing the boundaries of the Maya polities as they would have been during the early contact era. The principality around Xculoc [the village where I worked] is marked by the surname common to many of the villagers—Canul Pat. I am excited to show the map to Vitaliano, whose family has helped me so much. "Look this map is from 1579 and here's your family name, right here where Xculoc is. Your family has lived here for hundreds of years." He is excited as well, not because I'm showing him the map, but because the governor has promised to come to the village tomorrow. It later occurs to me that his family has lived here for hundreds of years,

6/2

Many of the little boys have
wrist rockets, which they use to
kill insects birds lizards (the latter
I've heard them tell, but have not seen)

I saw an incredible slice of reproductive
history today. A woman (mother) said she
was 76 & had a 8 yr. old son.
This I doubted until U. said she also
has a 50 yr. old child. 11 children
over a 40 yr. period. What a life.
She is also a sister of those bearing
the sons who die in preadolescence.
She has had 3 sons, 1 brother, who
died. Her youngest is 13 & my guess
doesn't have long to live. He has a
horrible deep cough. U. says "their
muscles go soft". This kids can't
walk. Still this woman doesn't
look 76. 65 at best. She agile &
very ambulatory. At 76 she's one of
the oldest & we've interviewed, she
doesn't seem it. When I get the
age of the oldest child from ego.
I'll check again. Can a & 63 get
pregnant? U. says there's a woman
70+ in Boleneben w/ a baby? I thought
only the bible had such stories....

Three days in Time was too long.
They get unused to us & I get unused
to them. Seems surreal that first
morning when you're barely awake walking
outside.

Spent a long morning in the fields,
observing two different women weeding.
Saw for the first time the gourds
that many use as canteens (chuu)
I couldn't think of the Spanish
word.

Walked past a field which had
fertilizer under the '3 weeks old plants.
It's supposed to make the corn grow faster.
Two types of corn of being grown this year
1. Hibride de blanco
2. Hibride de amarillo
   de blanco bears fruit sooner

I pointed out some insect to me,
looks like a small fat grasshopper, that
is considered a major plague ~ every
5-6 yrs. It's been 5 yrs. since
the last time. Insecticides can
diminish their effect.

Again the tractor is supposed to
Show up tomorrow.

Thinking about return rates some
more. It seems like there's little point
in getting several samples from one
person for each of these tasks. 1. Because

Sample journal pages exhibit how Maya village activities and personal observations are recorded in narrative form.

and evidence of that would not resound in the same way it does when we find a link to our geographically transient pasts.

The next day's entry reads:

> There is a great commotion in the village. A phalanx of broomed women sweeps the village literally from end to end. When I come back from spending the morning in the fields, the stone fences that line the village paths have been white-washed, the local analogue to rolling out the red carpet. The governor's entourage arrives and men and women assemble in the plaza, men on one side, women on the other wrapped in their robozos, wearing their best huipils and whatever shoes they might have. The governor in his speech bloviates that electricity will be brought to the village in the coming year, and a new water pump will be installed since the one they have is broken.

While neither happened, his speech alerted me to something I had overlooked. I had recently started living in the village and had not paid particular attention to the broken water pump. Every day after that I noted whether or not the gas-powered pump was working. At the time it was just another daily entry I made in my log of village events, and I was focused on a question about high fertility and children's time allocation, but I later became interested in energy expenditure and its effect on reproductive outcomes. When the gas-powered pump was built, women no longer had to draw by hand buckets of water from the fifty-meter deep well. This technology introduced a substantial savings both in women's time and energy budgets and is correlated with an increase in family size. When the well was functioning turned out to closely predict how hard women work. I had the time allocation data to reconstruct women's daily energy expenditure, but if I had not noted when the well was and was not working, I could not have made this link. I could not have anticipated that this detail, which seemed extraneous at the time, would later become fodder for a new research direction.

NEW CHALLENGES

A number of years later, I started a research project with the Pumé, a group of South American hunter-gatherers, in order to collect comparative demographic data among people who make their living in a very different way from the Maya. Much of what I had learned about field methods while working with the Maya was appropriate. However, there were also many new challenges.

As with the Maya, accurately aging individuals was essential to construct censuses, reproductive histories, and fertility and mortality profiles among the Pumé. In societies that keep vital records and mark the passage of time in calendar dates, this is relatively straightforward. However, births and deaths are not recorded in most small-scale societies, so recall and interview methods are necessary to age individuals. These difficulties are redoubled among the Pumé who, in addition to not keeping vital records, do not have a dating system and do not use names in the way we do. We were, however, able to establish accurate ages for most people using several methods, one of which was discovered quite by accident.

My husband has worked with the Pumé since the 1990s. When he initially walked into their camp twenty years ago, the first thing that happened was that village elders deliberated about how he would fit into their kinship system. It was decided that his tie would be through the village shaman as his older brother. Once this one classificatory relationship was established, relatedness to other group members followed from this affiliation. When I started living in the village many years later, I became related as sister-in-law. In this culture, these bonds are immediately established because everyday events cannot proceed without them. Who shares food with whom, how shelters are arranged, who is included in hunting and fishing trips, and who sits next to whom during rituals all follow from kin logic. Being tied into a kinship system sets up mutual social and economic responsibilities among members of the community and gives them a way to address each other. The Pumé use kin terms, not names, to refer to each other. This is common in many small-scale societies. Only when groups are composed of several hundred people do kin terms become insufficient to distinguish among individuals, and then names are regularly used. In addressing other villagers by kin affiliations, we learned the very specific terms that are used to refer to older and younger siblings. This gave us the necessary handle to establish birth orders, reproductive histories, and relative ages.

Anthropologists live in their study communities, often in very remote locations, with people who are intimately familiar with their environment and who are adept at surviving under conditions that are often harsh. In contrast, we anthropologists are ignorant about how to survive on our own—how to find food, water, firewood, or shelter. Because they live so successfully in their world, we expect them to readily explain the strategies that underlie the behaviors we observe. This can be trying, because from their point of view we are asking the obvious, a child's question. In seeking clarification, the response is often like one any of us would give to a child when we don't want to bother with a long and complicated answer.

For example, when the Pumé encounter hawks or other raptors at their nests, they shoot arrows at the birds and then climb the tree to throw the eggs out. They never use either the eggs or birds as food. My husband, who was studying Pumé subsistence strategies, thought they did this to reduce competition with raptors for the small game such as lizards, armadillos, and rabbits that are an important part of the Pumé diet. As was often the case when asked "Why do you do that?," the response was simply "because that's what we do." Adults know why. Because only a young child would ask such a question, the reply is appropriate to a child. It took two years of living with the Pumé for him to get the answer. He was on a hunting and root collection trip with a man and his wife when they came across a hawk's nest. The husband climbed up the tree and threw the eggs down to the ground. The wife turned and said, "We do this because the hawk people fly away and tell the deer people that we're out hunting them and scare away all the deer." This is an accurate description of hawk and deer behavior. Hawks set up an alarm call when the Pumé are about. Deer see very poorly, but hear very well. Deer flee at the sound of a hawk alarm call. This naturalistic explanation of their understanding of animal behavior and its influence on hunting tactics came when least expected.

ON NOTE-TAKING

It is difficult to anticipate all the needs your data will serve or the questions you will ask in the future. My advice to someone going into the field is to record everything you can, while you can, in as many different ways as you can. Although it is easy to assume that you will simply remember observations, come back later, or see something again, these may not happen given the vagaries of fieldwork. The ethnographic world is quickly transforming along with the rest of the natural world. While these changes are also of interest, the chances are few to document many phenomena before they are altered or modernized. It is a shame to have all the pieces of a puzzle to make an argument, but be missing a critical observation that you neglected to record at the time. There also is a less noble reason to document field experiences in varied ways. If you are living in the middle of nowhere, with no entertainment and with people who do not speak your language, musing about the world around you—whether it be through narrative, drawing, photography, or map-making—is a great companion and diversion. Connections to research often are made much later.

While repetitive observations are the sustenance of science, they can obscure connections and pigeonhole imagination. Understanding the scientific data we collect also requires being alert to clues about inter-relationships that are often outside the initial research problem. Narrative by nature is relational, and recording events, thoughts, speculation, and anecdotes as well as quantified data brings our curiosity back from the field. Somewhere in there is a story, a really good story that you will repeat time and time again. It was the bawdy tales of Margaret Mead, the gruesome stories of headhunters, the lavish descriptions of Plains Indians, and romantic narratives of living far from the beaten path that inspired me to become an anthropologist. Important connections are often made by accident, outside the bounds of our research agenda. How we record field notes opens or closes us to the unexpected.

7

In the Eye of the Beholder

JONATHAN KINGDON

THE HUMBLEST FIELD RECORD is always an act of translation. Whatever is recorded, whether animal behaving, plant yielding, dawn revealing: all have to be processed by human senses and translated into words, numbers, sketches, photographs, or any one of many other communicative conventions or devices that serve to inform other humans. The historical beginnings of those conventions may be recent and highly technical or take us way back into history or prehistory, but at the individual level every one of us learns the techniques and conventions of recording data from mentors, peers, or media such as this book.

Many of us first learn to make systematic field observations while at school or university, but in my case the beginnings of "note-taking in the field" began at a much earlier age. Because there was no school nearby and because my mother was a trained artist and art teacher, my first lessons from her were not in reading and writing but in drawing directly from life. I remember my mother sitting me down at the age of about five with pencil and paper to draw an acacia tree in the yard while she busied herself with her own sketchbook.

After a while she came over to see my efforts. "Splendid! But haven't you noticed how the trunk narrows as it rises? And see how the branches flatten out sideways, not like that oleander over there, where they all go up at a steep angle. Now don't rub that one out, just do another drawing to compare with the first one." My mother's injunctions to observe, accurately record, and compare things were, I only discovered very much later, the essence of scientific inquiry. Yet they were experiences that

Drawing of an acacia tree damaged by elephants in Amboseli, Kenya.

slipped by with no more significance than play: small but enjoyable details in the fabric of day-to-day existence on the shores of Lake Victoria
in the 1940s. There was one perhaps unexpected by-product of the sequence in which my schooling took place (drawing from life first, then
reading and writing, with math reserved for later, more formal classes):
verbal constructions and literacy became secondary to visual imagery in
my perception of the world around me and in my repertory of communicative skills.

After a conventional secondary schooling in a British boarding
school, I went on to a truly classical art education at the School of Drawing, founded by John Ruskin, at the University of Oxford. Here, under
the tutelage and examples of Percy Horton and Lawrence Toynbee, both
superb draughtsmen and passionate about their art, I not only refined
and diversified my skills in representation but expanded my cultural horizons with the knowledge that all imagery, my own no less than that
of the past, must reflect the preoccupations and values of an artist's

Four early drawings.

time and place. The school was housed within that university's Ashmolean Museum, and a mere stairway and corridor away was the "Print Room," a treasure-house where I was able, indeed encouraged, to study, copy, and even handle great Renaissance drawings by Leonardo da Vinci, Pisanello, and many others. I went on to a somewhat less rigorous program of training and experiences at the Royal College of Art in London but spent much time in the Victoria & Albert, Natural History, British, and other London museums.

My employment of drawings as an adjunct to scientific observation began with my desire to assemble what I conceived of as an "atlas of evolution" that doubled as a mammalian inventory for East Africa. At the time, around 1960, I was a young lecturer in the then University of East Africa, but had grown up having picnics in Olduvai Gorge, spending holiday weekends on the Serengeti plains (then a "game reserve"), climbing mountains, and exploring the shores and islands of East Africa's lakes and coast.

Sketches comparing
head-on views of a young
wart-hog *Phacochoerus
africanus* and a red river hog
Potamochoerus porcus.

I wanted to put together a project that would take me all over East Africa (what others might call "the field," but I conceived of as my homeland)—a project in which I could develop and articulate my already rich experience of mammals. I was also fascinated by fossils and fossicking and wanted to direct my energies toward a research project in what I knew must be its only intellectually satisfying context—a discourse on mammalian, including human, evolution.

The broader social context for my project was a continent just emerging from colonialism and the rigors of World War II. African colonies were becoming independent nations and new national parks were declared just as mass tourism began to overtake the handful of rich hunters who had monopolized the enjoyment of "Big Game" during colonial times. As well as tourists, energetic and curious young teachers, some with the Peace Corps, were appearing on the scene, and I met some of the first generation of foreign scientists that had begun to find their way to East Africa: American researchers under the Fulbright Program, Japanese primatologists from Kyoto University, and young British or European scientists, mostly working on doctorates. I was also fortunate in my resident colleagues in Kampala and Nairobi, including L. S. B. Leakey, and many other passionate local naturalists. I should stress, however, that apart from the disparate efforts of these pioneers, there had been no coherent program to study the richest mammalian fauna in the world, and not even the most rudimentary of field guides was available.

For me that represented a call to action, but an even more driving incentive was my own childlike but still unsatisfied curiosity about the "meaning" of shape or "form" in animals. What "shaped" animals? While I knew it was not some divine plan, as cooked up by Middle Eastern prophets under starlit desert skies, I wanted to achieve my own personal understanding of the shaping process through my own efforts and partly through my preferred medium—drawing. I had read Darwin's explanation of the evolutionary transformation of mammalian limbs and digits into hands, hooves, wings, and flippers, but I had also learned that many more exciting explanations and possibilities were concealed within the specifics of the animals' diverse biologies. Furthermore, an evolutionary inventory of my tropical homeland's mammals provided an academically sound vehicle for just such a quest. As a lecturer at the University of East Africa I was well placed to study my material and to secure the necessary permits.

So, taking the four Anglophone/Swahili-speaking countries of East Africa as a microcosm of the continent, my studies began with a typically Darwinian procedure in evolutionary biology—comparing the morphology of related species. I essayed into behavior, ecology, anatomy, and biogeography because, as I eventually explained in the preface to my book, "they may perhaps serve to increase awareness of the magnitude and magnificence of evolution."[1] I went on to point out:

[We] know that even slight dissimilarities in appearance between species can usually be related to functional differences in the way of life to which the respective species are adapted. In considering these formal dissimilarities, drawing seems to me to be, in its own way, as appropriate an expression of thought as mathematical formulae or tables . . . The comparison of forms raises questions and drawing can be employed as a wordless questioning of form; the pencil seeks to extract from the complex whole some limited coherent pattern that our eyes and minds can grasp. The probing pencil is like the dissecting scalpel, seeking to expose relevant structures that may not be immediately obvious and are certainly hidden from the shadowy world of the camera lens.[2]

The fieldwork, sustained for more than ten years, took me to the extremities of East Africa, from Kitgum in northern Uganda to Newala in southern Tanzania and from Bufumbira in central Africa to Kiwayu on the Kenya coast. During ten years of intensive research I drove over 150,000 miles in a Land Rover which was fitted with a winch and four boxes that slotted into the back well. One box was for bedding, another for food and cooking utensils, and the others with drawing and note-taking equipment as well as the paraphernalia of trapping and dissecting. I recorded my expeditions in a running log and made selective lists of animals I saw and, more comprehensively, almost all the mammals that were the focus of my current interest. If specimens were collected, they were preserved and conventional measurements recorded on labels.

I deliberately took no camera because I thought I might come to rely on it too much and therefore blunt the intensity of my observation. Notwithstanding this self-imposed limitation in the field, I later employed film, video, and stills to analyze gaits or details of ephemeral and sometimes very rapid sequences of behavior. In the years since my early studies, video and camera traps have become among the most useful of tools for the study of mammals, opening up many aspects of mammalian biology that were unimaginable in the 1960s. Simple outline drawings can be traced directly from photographs or film stills to illustrate a great variety of structures or behaviors, but such studio work hardly comes under the rubric of fieldwork. Preliminary or exploratory tracings can be drawn in pencil, but publishers tend to require crisp black-and-white line drawings. Depending on the scale or detail needed in the finished product, a felt-tip pigment liner (ideally waterproof on paper and light-fast) of diameter 0.05 to 0.2 drawn on 90g/m2-gsm smooth tracing paper allows tracings to be made to serve most purposes.

Photography and electronic monitors can be of enormous utility, but what the eye sees requires analysis in terms that are more commonly associated with an education in art than in conventional science. More specifically, it has been in my researches into the evolution of visual communication among mammals that I have found the techniques of visual analysis associated with the act of sketching or drawing from life to be most useful. The literature on visual communication often includes quantification of, say, the frequency of a display repertoire, and much useful information about the visual channel's role in a particular organism's biology can be revealed in this way. Nonetheless, any in-depth analysis of visual phenomena takes us toward a conscious awareness that there are many visual phenomena that have evolved, through natural selection, "in the eye of the beholder." I will return to this truism later while describing my studies of head-flagging in guenon monkeys of the genera *Cercopithecus, Allochrocebus, Chlorocebus, Miopithecus,* and so on. With broader fields of interest and less time to squander, I now tend to go into the field with a digital camera, and I am a lot less of a purist about diluting my observational skills.

Back in the 1950s and 1960s my principal tools were a sketchbook or pad of good quality white paper, a penknife, and a B pencil: B because so long as it can be kept sharp, every tone and the finest detail can be recorded without having to change pencil. To make notes about color I try to keep in reserve a fabric roll with multiple slots to hold a spectrum of watercolor pencils. Color notes are most rapidly and easily made with the help of these pencils, but whenever they're not immediately to hand I often find myself resorting to written notes on color along the margins of a drawing. My travels, particularly during the wet season and at a time when virtually no roads were paved, sometimes ended in an elephant-made quagmire, on the edge of a raging flash-flood, or with my vehicle in sand, mud, or deep water. A winch was most useful for extricating myself, and it also helped me skin and manipulate the carcasses of large animals when I took advantage of the local government "game control" or "cropping schemes" that were in vogue at the time.

There was no lack of other obstacles to overcome, and occasionally vigilante villagers could be aggressive, but my knowledge of Swahili and local customs usually calmed tempers, and I always made a point of introducing myself to local authorities before starting work in a new locality. Children and hunters were always intensely curious and sometimes brought in small game for me to see. Sometimes I could surprise them by pointing out some functional detail or facet of the animal's biology, and sometimes they surprised me with their own knowledge or fables. I usually preferred to sleep in my hard-body, long-wheel-base Land Rover to avoid foraging army ants, scorpions, thieves, or large animals tripping over guy-ropes at night, which could make camping a hazard in some localities. Swarms of insects, particularly mosquitoes, sweat-bees, and flies (tsetses, bluebottles, bot- and horse-flies, and plain old house flies) often made drawing difficult as they bit or covered my paper, hands, or face and clogged my eyes, nostrils, or mouth while I tried to draw. Lemongrass or other repellants helped, but I discovered that it takes a lot to discourage the more determined of invertebrates! When I was called to the corpse of a dead animal, especially when it had been dead for a while in the hot season, the stench brought avalanches of flies and flocks of vultures, and this dissecting called for a strong stomach. My drawing of one dissected rhinoceros head was done under such stinking conditions that my tough Masai assistant actually threw up during that particular dissection!

In likening my pencil to a scalpel I was, of course, stating my explicit interest in anatomy, but in my mind dissection was more of a physical metaphor for "revealing the hidden," and this was a major incentive for embarking on my mammal studies. Early on in my project I was sent a fetal ground pangolin and I discovered that the precise geometry that emerged in my drawing was far less apparent in the coat of abraded scales of an adult. In another drawing of a flayed tree pangolin, I found a wealth of functional details hidden beneath the animal's scales and skin: the cartilaginous remnants of ear pinnae; a soft, fingertiplike terminus to the tail; powerful claws, digits, sinews, and muscles; and enlarged subcutaneous muscles (above which the flank scales are embedded, serving to protect the pangolin's abdomen).

At the time I compared the anatomy of yet another pangolin, the giant species, and in making many drawings of the three pangolin species, all very poorly known, I came to conclusions about the relationship

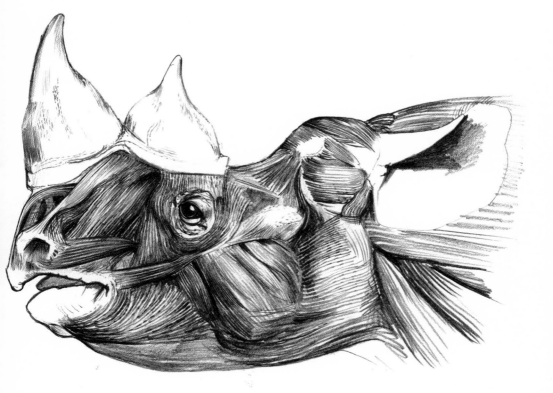

Dissection drawing of a rhinoceros *Diceros* head.

between pelvis structure and locomotion differences—quadrupedal, bipedal, and arboreal—that I subsequently published and that had not previously been remarked upon or discussed.

In the age of instant digital photography it may seem perversely old-fashioned to put a value on the slow, primitive, and inaccurate techniques of manual drawing. Photography teaches us that the very act of putting a line around the edge of an observed object is an artifice. Such outlines rarely appear in photographs, or, for that matter, in nature, and yet . . . and yet? Contemporary research on the human brain shows that it does NOT process images as a neutral camera does. The brain finds edges and builds constructions that are at least partially based on previous experience—possibly including past contacts with artifacts such as "drawings" as well as previous knowledge of natural objects. Visual neurobiology is a discipline in its infancy, but it confirms that visual constructions are both

Sketches of fetal
and adult ground
pangolins *Smutsia
temminckii*.

Dissection drawing of a tree pangolin, *Phataginus tricuspis*.

complex and integral to cognitive development. This implies that even an outline sketch that bears little relationship to the so-called objectivity of a photograph might actually transmit information to another human being more selectively, sometimes even more usefully, than a photograph. For example, a few quick sketches of a hippopotamus allow the difference between sexes, the peculiar architecture of amphibious existence in a giant quadruped, and the combination of biting and antlerlike clashing of enlarged lower jaws to be appreciated at a glance.

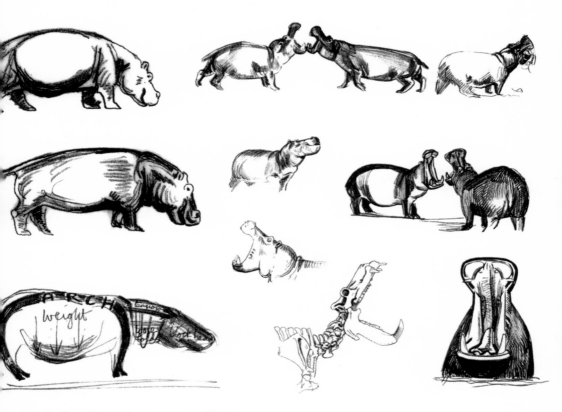

Sheet sketches of hippopotamus anatomy and behavior.

If the brain is unlike a camera in actively seeking outlines, there is the strong implication that "outline drawings" (just to take a single type of visual expression) can represent, in themselves, artifacts that may correspond more closely with what the brain seeks than the charts of light-fall that photographs represent. Field observations not only have to select out data that are relevant to the questions being asked, they always translate perceptions into some medium other than that in which they were originally transmitted to the primary viewer.

Drawing therefore represents a species of translation that is different from what emerges in photography. Given the new research on how the brain processes visual input and given that drawing is a mental process, no further justification need be made for the utility of drawing in lifting out relevance from within the chaos of actual visual experience. Visual experience being extremely diverse, it is inevitable that drawing, too, is a very diversified skill. Learning to discriminate between what is significant and what is irrelevant to the question at hand is an essential part of field studies, and just such discrimination is integral to the act of drawing.

It can be credibly argued that all drawings carry cultural and psychological baggage. We recognize that a drawing by Leonardo da Vinci, by Katsushika Hokusai, or by Pablo Picasso requires cultural knowledge for the fullest interpretation, but even in the absence of a translator all are accessible to comprehension, even by visual illiterates. A cat drawn by any one of these masters can be more easily referred to that felid than the three letters C, A, and T, for which the English language is an essential predicate. Likewise, if *Felis catus* is to be distinguished from *Panthera tigris*, Latin for a tiger, by reference, say, to its weight, the artifact of our chart or histogram employs the further convention of Arabic numerals. Even scientific terms and procedures carry some cultural baggage!

Such quibbles aside, the mental activity that is uniquely associated with the act of drawing is, quite literally, an act of "figuring" and looking at drawings can be an active retracing of this figuring process. Recording and interpreting things that are seen not only involves a different medium of cognition, but the "notes" that emerge can often be a lot faster to record than their clumsy verbal equivalents.

In my case I set out to employ my skills and find various graphic ways of expressing my own explorations of multifarious evolutionary processes on home ground in Africa. Furthermore, on leaving my European schooling behind and returning to East Africa I judged evolutionary biology to be the most exciting, truthful, and intellectually challenging expression of my time and culture. So I focused in particular on what could be learned by closely examining and drawing humans and other mammals from a deliberately Darwinian viewpoint, with the working of natural selection a major preoccupation.

One lesson that I was quick to learn was that the eyes of viewers have been a primary selective agency in determining the external appearance of many animals. Any visually oriented animal needing to discriminate between friend, foe, or mate among their own species must judge, select, be selected, signal, or slip away on the basis of appearances. The caracal cat is a fine example of an inconspicuous animal that uses small movements of its head and black-tufted ears to flag information to another caracal. The slightest twitch of its ear-tips can signal mood, status, and intentions in a manner analogous to flags on sailing ship masts.

The iconography of caracal ear- or head-flagging is intricately crafted, and fingers on a pencil can scarcely keep up with the rapidity of their flickering movements. Nonetheless, I believe drawings can be a clearer medium for exploring such a visual Morse code than laborious written accounts or quantified records of frequencies. Future students will, no doubt, use new photographic sources for further analysis, but when I did my sketches, I felt that it was enough simply to observe and advertise the existence of such interesting behavior. The interest of tufted ear-tips has become even greater with the recent discovery by molecular scientists that the caracal is not a close relative of northern lynxes, which have independently evolved similar ear tufts. Such convergences in other cats, primates, squirrels, and some antelope species reinforce our awareness of similarities in function between our own hoisting of flags and the ear-tip devices used by other animals to communicate.

Artists and scientifically minded humans are not the only animals that seek to lift out significant or informative "form" from the chaos of nature. To survive, every visual predator, whether cat, hawk, or tiger-fish, must repeatedly "see through" the disguises used by their prey. Because visual predators often select the most easily seen individuals of their prey, the evolutionary explanation for progressively better and better camouflage in prey animals lies in processions of survivors and offspring of survivors that were somewhat less easily seen and caught. The visual acuity of predators has therefore been a primary agency in their preys' external appearances. Thus concealing coloration is as much a manifestation of predator behavior as it is that of the prey. Furthermore, because every prey animal lives within a setting where it will be seen or escape

Conspicuous white
vertical becomes
narrow horizontal

← cuts off here
when back

Caracals ear well designed
to serve as signal as tip
is extended both front or back
line can be major vis component
black base to ear is triangle that
is "cut off" when ear is back (seen from side or ba

nervous yawn

black markings
point or focus
on eye

to show how tip can extend front or back

when ears back
throw all visual emphasis
back & creates a tension
between front & back

Drawings of a caracal
cat *Caracal* flagging
its ears and head.

shortening of temple
in alert

when at all
nervous flags head
from side to side
with ears tending
to flat & eyes closed
a bit

...gation of temple
... with ear back

...ngs create
...es linked by
...f slightly differentiated hair
...permutations very numerous

Drawing of a checkered
sengi *Rhynchocyon
cirnei reichardi.*

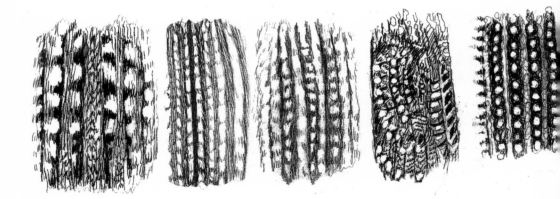

Drawings of cryptic patterns in (*left to right*): *Rhynchocyon cirnei reichardi*
(Afrotherian checkered sengi); *Lemniscomys macculus* (African grass mouse);
Spermophilus tridecemlineatus (American ground squirrel); *Caprimulgus
pectoralis* (African nightjar bird); and *Cnemidophorus* sp. (whiptail lizard).

being seen, its appearance will have been selected in direct relationship to some specifics of that setting. We call the outcome camouflage, but it is actually a striking manifestation of appearances being translated into another medium. In effect, predator selection "draws" or "paints" some aspect of the landscape onto the bodies of surviving prey animals. Hair, feathers, or chitin become the medium for miniature landscape paintings.

Studying how camouflage hides an animal is the precise opposite to lifting its profile out of its setting, but it is no less a fascinating study for the analytical viewer and as much a subject for drawing as any other manifestation of evolution. Perhaps it is even more so, since drawing is little more than tonal scratches on a surface and the coats of animals are made up of innumerable scratchy hairs of different tones. Camouflage is found on both predators and prey and there are at least two distinct classes of "camouflage." One, usually very small-scale (and commonest in arthropods and marine organisms), consists of patterns that match the animal's or plant's immediate substrate exactly. The other, commonly larger-scale, consists of abstract patterns in which the disposition of light falling on broken ground or the chaos of plant growth is mimicked by patterns that average out the relative proportions, disposition, and shape of a limited number of tones.

Averages, by definition, are not specific, so it becomes very interesting to find animals from different classes sharing patterns with a common averaging-out of three or more tones in patterns of light-colored "dots" on semilinear dark "lines" with intermediate tones in between. In this respect it was not enough for me to faithfully record just such a coat pattern on the checkered sengi, or elephant shrew *Rhynchocyon cirnei reichardi*. I have used the simplest of "scratchy" drawing to explore this and other examples of predator-selected patterns in rodents, a bird, and a lizard. The outcome of such carpetlike drawing could hardly be more different from the representation of, say, facial expressions or the outlines of an active mammal, but all express hands at the service of a curious mind and an attentive eye.

Drawing has long been an adjunct of anatomy and illustration, but I wanted my studies of mammals to take the practice further: my "wordless questioning of form" had to document activities such as feeding, copulating, excreting, fighting, and signaling information to conspecifics. It was through the process of watching and drawing such vital activities that I came to articulate an understanding of "shape" that fed back into what might even be called "drawing with Darwinian perspective." However, there was a mismatch between my ability to collect, measure, and interpret the physical, sometimes dead, products of evolution in a detached, scientific spirit and my personal appreciation that mammals were as alive as I was and as full of "moments" in the pursuit of their lives as I was in mine. This was nowhere more apparent than in my struggle to marry the fleeting moments of ethology, or "behavior," to the more lasting "solids" of morphology. For me, drawing was an essential medium and adjunct for trying to bridge that divide.

To think that morphology somehow "constrains" or governs behavior is to invert the evolutionary reality. Indeed it blurs our insight into the opportunistic nature of natural selection itself. Instead, it is one of Darwinian evolution's most profound insights that morphology has emerged incrementally as the outcome of individual and varying animals probing all the possibilities of their lineage's existence under the limitations of short lives lived out in specific localities. Over evolutionary time, and to a significant extent, natural selection of small details of individual behavior would seem to have driven and shaped morphology, not the other way around. Those individuals with behaviors appropriate to their setting and time have been the most likely to survive. Riding on the coattails of such behaviors are those small physical and physiological differences that may further enhance the effectiveness of the behaviors. We know that, incrementally, such physical changes can reorder the shape of beaks, ears, limbs, even entire body proportions. In the process, species emerge. By helping me lift out those details of structure that define one species from another, drawings illustrated my books. But in addition, the process of making them was integral to discovering more about the animals' biology and evolution.

In the previous paragraph I used "probing possibilities" as a metaphor for the dynamic outcomes of natural selection, but "probing" becomes material and physical when an evolving lineage of mammals goes underground in pursuit of food or shelter. Among Africa's most ancient mammals, the Afrotheres (which include aardvarks, elephants, sea-cows, and tree hyraxes), are many species of golden moles. At first glance, or in a photograph, golden moles look like animated turds in spite of the pretty metallic fur that gives them their name.

Closer inspection reveals a leathery spade-shaped snout at the front end and tiny clawed feet that row them along without being

Sketches of a golden mole *Chrysochloris stuhlmanni.*

able to lift their bellies off the ground. Such "shapeless" animals raise problems of representation. Yet, in spite of being wholly enveloped in a chador of fur, golden moles are no less highly evolved than elephants or aardvarks. More penetrating study reveals a seamless convergence between "probing" and the forms that have evolved to serve probing or digging behavior. Thus a digging golden mole forces its nose, which has become wedge-shaped, deep into the soil. Then a finger, armed with a curved, pointed claw, is brought forward into the nose-made crevice, and a very powerful opening-up action follows as the robust skull pushes up and the claw tears down and back. At the time of my studies I drew two simplified anatomical diagrams to illustrate this action, but I also made a quick sketch of the little cadaver that lay in the palm of my hands. It seemed to me that its lifeless form contained within it the tiny digger's vivacity as well as the abstract forces of natural selection that had shaped the survival of its ancient ancestral line. For me, that simple sketch

summarizes in a way that descriptions cannot how the behavior of probing and digging has selected for what we can only very inadequately
describe as a "streamlined" skull and body shape.

"X-ray" sketch and two diagrammatic dissection drawings of a digging golden mole
Chrysochloris stuhlmanni.

Another example of the potential for drawing to lead to original and
rather surprising results emerged from a 1966 attempt at primate por-
traiture. Long before then, scientists, noting differences in the length of
primate muzzles, had quite reasonably posited that long-muzzled terrestrial baboons living in open country had evolved from short-muzzled
forest-dwelling ancestors living in the trees. With the passage of time
this generalization hardened into an orthodoxy whereby the baboons'
ancestral "descent from the trees," as in human ancestry, came to represent or imply progress among the primates. Any reversal of this stereo-
type seemed unsettling, and I too was momentarily disconcerted when
I found myself obliged to contradict what had come to be presented as
a directional principle by some authorities. Drawing the faces of short-
snouted and mainly arboreal mangabeys and comparing their skulls
with those of other primates, I found mangabey facial bones had been
"dragged" into peculiarly warped and wrinkled shapes that could only be

explained in terms of a backward migration of the muzzle in their ancestors.

Furthermore, there was nothing "primitive" about these baboonlike monkeys, and their habits and habitats were more consistent with an evolutionary "return" to the forest than with their being primordially arboreal. The idea that much of the hierarchical male behavior associated with elongated muzzles had become obsolete with secondary arborealism was consistent with my demonstration of cranial evidence for a secondary "retraction" in the size and length of male mangabey muzzles. For me the wrinkled bone and "buckled" cheeks of mangabeys seem to illustrate how behavior, modified in response to changing environments, then leads on to morphological change.

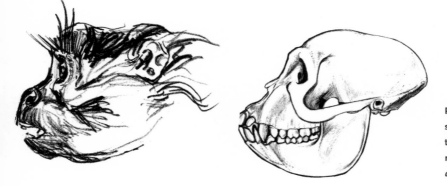

Profile sketch and skull drawing of the gray-cheeked mangabey *Lophocebus albigena.*

The idea that mangabeys are derived still has its skeptics, but it may be significant that none of the critics, so far as I know, has ever attempted to sit down and draw mangabey skulls and faces, let alone compare the outcome with a dozen or more related primates! To explore just how mangabey skulls had been remodeled I adapted an old artists' trick. Laying a symmetrical grid over an outline drawing (in this case, a long-muzzled monkey skull), the grid's points of intersection were then plotted over another outline drawing (in this case a mangabey skull). The resulting grid lines followed exactly the contours of the mangabey's buckled cheekbones, a remodeling of skull components not dissimilar to that commonly found in some whales and bats.

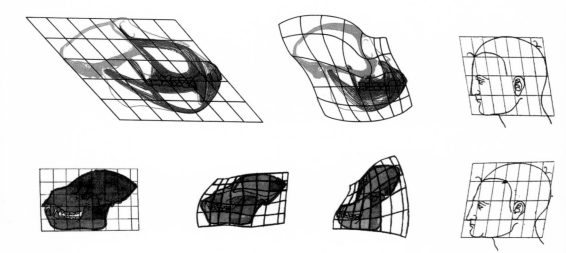

"Cartesian coordinates" imposed on skulls. *First row, left and middle:* a long-faced monkey (*Papio anubis*) and baboon mangabey (*Lophocebus albigena*) illustrate hypothetical phylogenetic retraction in mangabey skull. *Second row:* Fossil monkeys: *Aegyptopithecus; Afropithecus; Sivapithecus.* At far right, two human faces on a coordinate system from Albrecht Dürer's sketchbooks of 1514.

This use of coordinate grids emerged in the 1960s when I took my students to make drawings from dissected apes and humans in the Makerere Medical School in Uganda. This was encouraged and facilitated by colleagues at Makerere University, particularly Alan Walker (now at Pennsylvania State University), an accomplished artist in his own right, and Clifford Jolly (now at New York University). Both were profoundly influential in their very wide-ranging but at the same time disciplined and focused studies of the evolutionary process. Alan possessed an astonishing ability to correlate the detailed structure of living and fossil primates with the nuances of behavior to which they are adapted, and Cliff conducted field studies that linked behavior and social organization with morphology. We introduced Makerere students to D'Arcy Wentworth Thompson's *On Growth & Form* (1942) and his Cartesian coordinates, a visual technique for comparing one form with another (which in turn had been resuscitated from Albrecht Dürer's sketchbooks of 1514). It was these coordinates that I employed to illustrate the evolution of mangabey skulls.

Alan and I also pointed out the correlation between upright stance and the relative size and orientation of the gluteus, or buttock muscles, noting that these shapely bulges are major actors in the task of holding and balancing the human body upright. I made drawings of this diagnostic area of anatomy from both human and ape cadavers in which the apes' gluteus showed up as very poorly developed but with impressive endowments in testicles (and, in a female, with sexual swellings around the vulva).

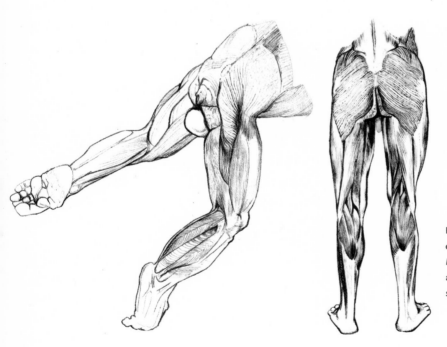

Legs and gluteus muscles of a chimpanzee *Pan troglodytes* (*left*) and a human *Homo sapiens* (*right*).

All this led on to many questions and conjectures, some of them humorous, some ribald, but in the midst of all these comparisons I found myself returning, again and again, to an old and unanswered question. What sort of behaviors and ecological settings might have driven the switch from four legs to two legs?

It was studying foraging apes and monkeys in Uganda and, much later, watching a field-study group of (mainly English) students of ecology collecting microfauna on the floor of an eastern Tanzanian forest that confirmed for me this point: it was the utility of hands as versatile food-collecting devices that must have initiated and driven the divergence between apes and hominins. Followed up with a wide-ranging reexamination of hominid anatomy, still more comparisons (using Cartesian coordinates again), and still more drawing, the work eventually culminated in my recent book *Lowly Origin*. Once again, exploratory drawings of bones and muscles were an essential part of the conceptual process. Developing one of Cliff Jolly's most influential ideas about the role of "squat-feeding," I eventually synthesized some radical theoretical innovations which have been graced with the rather unflattering title of "the bum-shuffling hypothesis."

Watching and drawing primates in Uganda was a starting point that led me to attempt an analysis of those patterns and colors that appeared to be used by monkeys to communicate with one another. In engaging with the difficult question of how visual signals are used and how they evolve, I found myself wandering into a no-man's-land between art and science. After all, the human primate uses visual signals in abundance, some of which get to be labeled as art. As for artists, throughout history they have been quite analytical about our capacity for representation and the many questions that the visual sense poses for us.

The red-tailed monkeys that set off my quest had been familiar to me from early childhood, and their expressive red tails and coiffeured, masklike faces had always fascinated me. A sheet of quick sketches from 1967 gives an impression of their protean form but cannot capture the rapidity of movements, especially of their heads, which are sometimes shaken or woven in what I first took to be efforts to shake off a persistent fly!

In order to familiarize myself with the details of face patterns, I drew very detailed frontal portrait and profile views of captives. Since

this required an exceptionally active animal to be still, I employed the drug Cernilan, injected into the tail, to temporarily sedate my subjects. A colleague who was investigating yellow fever epidemiology in western Uganda also supplied me with primate cadavers, and once I began to compare my red-tails with a still wider range of species, I had to resort to museum skins taken from remote corners of the Congo basin, the Niger Delta, or the hinterlands of Sierra Leone.

It soon turned out that the red-tails belong to a complex of very closely related, allopatric species (the cephus monkey, or *Cercopithecus* (*cephus*) species group) that range through the forests of equatorial Africa. Although they are similar in size, behavior, general body-build and color, their face masks differ from species to species more than they differ from any other monkey. Why would these, the most flamboyantly "masked" of all monkeys, be so diverse in this one feature while otherwise remaining relatively dull and uniform?

As with any evolutionary question, I had to begin by identifying a likely selective agency. The selectors for an intricate face-mask had to be members of the same species, and the context could at least begin with the two polarities of social organization—behav-

Montage of drawings exploring postures, proportions, and details of anatomy pertaining to "squat-foraging" in both extant and hypothetical ancestral Hominins.

Sheet of sketches of the red-tailed guenon monkey *Cercopithecus (cephus) ascanius shmidti.*

iors that serve centripetal cohesion or their opposite, centrifugal dispersion.[3] In monkeys this tends to translate into the antitheses of appeasing or aggressive behavior. In the case of guenons, dispersion is mediated by aggression (mainly, but by no means exclusively, male). Inasmuch as the face is involved, the aggressor stares head-on, grimaces, and attempts to bite or chase his or her opponent. The appeaser, by contrast, ostentatiously avoids eye contact, shyly looking down, up, or sideways, appearing to look at anything or anywhere to avoid staring at the opponent.[4] Here was the clue! When I watched them closely (and eventually filmed them), the

Profile drawings of
Moloney's gentle
monkey *C. (nictitans)*
moloneyi (above) and
C. (cephus) ascanius
shmidti (below).

headshakes that looked like fly-avoidance were actually very fast eye
eversions. I reasoned that diverting attention onto an eye-catching
"flag" could serve social purposes because conciliatory behavior favored cohesion. I went on to argue that the evolutionary beginnings
of such flags lay in the need to distract viewers from the eyes of

. So a side panel makes sense when head turning is ritualised

♀ puts head down when shy & submissive

In this posture the animal is very inconspicuous

♀ seen presenting her (cheek for grooming to) a juvenile

Impact of moustachios varies alot as there is much individual variation in the shape & extent of white & intensity of blue (very pale whitey b— in some cases & variation in the black bristly area.

Note that such a tiny structure as a nostril is difficult to advertise

Nose-sniffing in cephus same as in other spp of Cercopithecus possibly more frequent in all certainly in so.

The white marks are aligned with the nostril slits & may serve to advertise them so cephus could be a ritualised nose-sniffer

cephus

The ears are bare blue conches (but duller than face) The yellow cheeks are separate Chest tinted-bloomed with blue as is scrotum BUT all lost to distant view by grey fur

Black temporal streak links eyes & ear movements

yellow flush   the 3 zone   yell flush

Low black lower cheek mark ends at same point where yellow zone ends. i.e. point where frontal angle loses signal value

mouth

Sheet sketch of annotated quick sketches of cephus monkeys.

the viewed! The precise design of the flag was less important than its power to eclipse any potential ambiguity in the facial expressions of the signaler. The most plausible explanation for the evolution of such different masks is that, by an accident of climate change, sustained drought isolated regional populations at some early stage in the differentiation of face patterns, after which selection favored facial flags with ever-greater "power to distract." This was achieved by ever more geometric designs and ever-stronger contrasts in arbitrary colors and tonality.

Among the users of this highly ritualized headshake were males attempting to approach females. So here was a second clue to selection for face patterns—an adjunct to successful reproduction! In an environment where females live much of their lives scattered through a lattice of branches and twigs, it is particularly vital that, with countless avenues of escape, flighty females can be approached by potential suitors. "Presenting" genitalia is the commonest invitation to contact and clearest gesture of conciliation in terrestrial monkeys, but it is a lot less practicable during the fast life of cephus monkeys on canopy branchlets. In effect, the flag or masklike patterns in cephus monkeys that advertise eye eversions represent an evolutionary transition of the main conciliatory signal from the hind end of the body to the front. Significantly, cephus monkeys have inconspicuous genitalia.

Once again, this study shows how a diagrammatic drawing can put across the visual significance of such inversion of a signal more effectively than can text. When this piece of research was filmed I made a lifelike monkey model with a colorless face. The camera tracked me seizing the model by its genitals, whereupon I twisted off its bright blue scrotum and immediately transferred it to the face, where its masklike structure closely resembled the blue-faced mustached monkey, a close relative of the red-tail. An evolutionary development that might have taken millions of years to evolve and required a lengthy text to describe was symbolized on television by one swift, vulgar, but eye-catching gesture! Annotated sketches and diagrams made during the course of this study may help illustrate how direct observation of the behavior and morphology of these monkeys allowed me to assemble the evidence for an understanding of the evolution of their extraordinary masquerade. The drawings were made from wild monkeys in the forest, from captives,

Data sheet showing behavioral tallies for *Cercopithecus ascanius*.

and from specimens in museums, but all of them, no matter how detailed or how cursory, are essentially translations—translations from "the field" into the multifaceted but essentially experimental and tentative language of evolutionary biology.

The act of drawing serves to remind us that hands are agents of thought and experiment. The great cave drawings of Africa and Europe attest to that truth no less than the sketchbooks of Leonardo da Vinci. Photography has a great future, especially in the hands of imaginative scientists, but no matter how much ancillary wizardry photography accumulates, it will not be in competition with "drawing" in the broadest sense of that term. There will always be a role for exploration by the hands, encumbered by no more than a piece of ocher or a stick of charcoal. It may not be a medium for all, but its very simplicity commends it to the field biologist. Drawing therefore has a future as well as a venerable history. Its expressions have the potential to leap across great divides of time and place, but its practical utility is as a manifestation of the mind struggling with the meaning of what it encounters and what it wants to explore.

8

Why Sketch?

JENNY KELLER

And you who claim to demonstrate by words the shape of man from every aspect of his membral attitudes, dismiss such an idea, because the more minutely you describe, the more you will confuse the mind of the reader and the more you will lead him away from a knowledge of the thing described. Therefore it is necessary both to illustrate and to describe.

—*Leonardo da Vinci*

Another of my occupations was collecting animals of all classes, briefly describing and roughly dissecting many of the marine ones; but from not being able to draw, and from not having sufficient anatomical knowledge, a great pile of MS. which I made during the voyage has proved almost useless.

—*Charles Darwin, "Autobiography"*

FROM DA VINCI TO DARWIN, drawing has a long and illustrious history as a means of scientific investigation and communication. In this chapter, I hope to make it clear that this practice still maintains its relevance to scientists and naturalists. Although technological innovations have provided powerful new tools for documenting information, all field scientists can benefit from understanding how to think visually and can use simple drawing techniques to improve the way that they document their corner of the natural world.

DRAWING TO OBSERVE

Why should your field notes include drawings? For one thing, drawing makes you look more carefully at your subject. As an observational tool, drawing requires that you pay attention to every detail—even the seemingly unimportant ones. In creating an image (no matter how skillfully), the lines and tones on the paper provide ongoing feedback as to what you have observed closely and what you have not. If, for example, up to a certain point you have ignored the toes of your mammal, a quick glance at the toeless creature on your page will direct your attention to precisely that neglected feature. Just the act of making the drawing will force you to examine each and every part of your subject.

Okay, no doubt some of you are thinking at this point, "Well, sounds good, but I don't know how to draw." To these readers, especially, I maintain that one needn't be good at drawing in order to take valuable visual notes. Information-rich sketches can be achieved by anyone willing to put a modest amount of effort into acquiring some basic skills. In fact, there are some very effective "drawing" and color-recording techniques that require no training at all. For those interested in experimenting, some of these techniques will be discussed later in this chapter.

As it enhances observation, the process of drawing can also reveal different, unexpected aspects of a subject under study. Once when working with the late, esteemed marine biologist Ken Norris, I was asked to create a short animation of the aerial spin and splash of a Hawaiian spinner dolphin. At that time this process was accomplished entirely by hand, so many hours were spent in front of a video screen counting off frames in order to capture the proper poses and timing. By the time I was finished, I knew more than I had ever wanted to know about the length of each phase of the dolphin's spin and splash. In a casual comment to Ken one day, I observed that the splash-and-bubble phase lasted ten times as long as everything else. I was really only complaining about having to draw so many boring bubbles, but to my surprise, he was quite excited at this news. Apparently, it added quantitative evidence to support his theory on the importance of the bubble trails in dolphin communication. In science illustration as well as in science, you never know what will turn out to be important.

Sketches created while in the field can also record valuable information—sometimes even more reliably than photography. Although cameras are indispensable for capturing fleeting events and complex detail (and I would not go into the field without one), they cannot do everything. Colors in photographs are typically (sometimes dramatically) inaccurate, proportions are often distorted, and key features of the species may not be recorded clearly (or captured at all). Use of a camera can impart a false sense of security, as well, especially when a quick check of the digital screen seems to show us a perfect likeness of our subject. It is only later that we may discover that something crucial is missing: there is no view of the underside of the leaves, for example, or the animal's tail does not appear in any of the photos.

On the other hand, a simple image drawn on the page provides a perfect framework on which to record (and assess the thoroughness of) our observations. Basic shapes, arrows, circles, spots of color, and written notes can effectively document important field marks. A single line may be used to describe the arc of a hummingbird's dive, or the angle of a bird's central axis as it perches. A set of numbered, dashed lines on a page could even map out the paths taken by predators and prey in a chase across a meadow—a sequence that might be cumbersome to describe in words. In the case of a drawing I made of predatory tunicates, there were subtle changes in their transparent bodies that my camera could not capture but my sketches could show clearly.

A final reason to sketch while taking field notes is to get yourself thinking visually about the publishing stage of your research. Even if you plan to hire a professional artist, the illustrations will turn out better if you have your ideas clearly in mind beforehand. Of course, those well-crafted images will also help your written presentation convey the professionalism, significance, and interest of your work. It has been noted that scientists as well as laypeople typically begin their perusal of an article by reading the captions and looking at the pictures. If you remain unconvinced of the importance of illustrations in your work, heed the advice of professionals working in the field of scientific publication. Scott L. Montgomery in his book *The Chicago Guide to Communicating Science* suggests, "the visual dimension to science forms a language all its own, a kind of pictorial rhetoric, if you will. By this I mean that graphics are often much more than a mere handmaiden to writing. They don't just restate the data or reduce the need for prose, but offer a kind of separate 'text' for reading and interpretation."[1] The senior art director at *Scientific American*, Edward Bell, told me that he and his staff strongly encourage contributing scientists and authors to create rough sketches of the illustrations that will accompany their articles: "Nothing is better

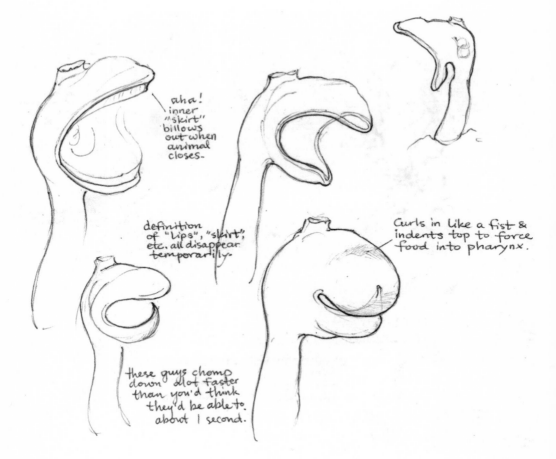

In the drawing, the following handwritten annotations appear:

aha! inner "skirt" billows out when animal closes.

definition of "lips", "skirt", etc. all disappear temporarily.

Curls in like a fist & indents top to force food into pharynx.

these guys chomp down alot faster than you'd think they'd be able to. about 1 second.

Subtle changes in the transparent bodies of predatory tunicates during the movement of their incurrent siphon (the open "mouth") are captured in these simple ballpoint pen drawings.

for the artist than some sketches from the writer's hand. Even if they are really crude and rough."[2] *Scientific American*'s assistant art director, Lucy Reading-Ikanda, added: "The biggest problem for us is if the scientist doesn't have a clear vision of what they actually want to show. Because the art is what people peruse first, it's what they 'read' first. *It tells the story*" (emphasis in original).[3]

The time to think about how your work will be portrayed is while you are in the midst of doing it. As you are observing subjects in the field, be on the alert for visuals that will help explain your story. Even if you don't draw, try at least to jot down ideas of how a particular topic might be illustrated, in preparation for eventual brainstorming with an illustrator. Once while working with an invertebrate zoologist on a scientific paper, it became obvious to both myself and the scientist that the roughest of field sketches would have made our job easier and the results more precise. The creatures I was being asked to illustrate, placozoans, are small (~1 mm in diameter), thin, ciliated animals that continually change their external shape. In the scientist's field notes, she had written brief descriptions of the different shapes they assumed: flattened, arched, wavy, cupped. These words seemed more than adequate to describe the morphology of animals that basically look like tiny pancakes. As the illustrator, though, I needed to find out more. "Do you mean wavy like a potato chip, or wavy like a washboard? And was that one cupped like a contact lens, or cupped like a cereal bowl, or cupped like the ballooned-out dome of a jellyfish?" My questions were playful but sincere—an illustration can only show specific shapes, not generalities. Once a line is placed, it conveys that particular edge and no other. I knew I would get back to my drawing table and have to put those lines somewhere, and I wanted the person who had observed the real, live animals to determine just where those lines would go. The scientist reviewed her notes, searching for further clues, and then sighed. "Wow, I wish I had more to tell you." In the end we felt safe in choosing mid-range positions for the illustrated animals. At the same time, though, we were struck by how little information the verbal descriptions actually conveyed, and how easy it would have been to include a few wavy lines, arcs, and ovals in the margins of the notebook.

At some point in every discussion about scientific illustrations created for publication, the subject of photography arises again. A scientist might say, "Okay, it might be handy to add sketches to my notes while I'm in the field, but why would I need illustrations for my final publication? For that I'll definitely use a photograph." The fact is, scientific illustrations can achieve certain things that photographs cannot. A good illustration can portray difficult-to-photograph or rarely witnessed events. It can incorporate everything that's important into one single image or show a special view of a subject—something cut away, exploded, turned semi-transparent, and so on. In a good illustration, you can create a representative "average" or "typical" specimen from pictures of separate individuals, or emphasize only the most important information about a subject, leaving out distracting clutter. A useful illustration might go back in time to show extinct species and scenes of the past, or it might portray scenes of the future or phenomena that do not yet exist. Think of it. None of us would ever have seen a picture of a dinosaur, or the topography of the ocean floor, or a cutaway of the sun, if not for illustrations. Likewise, it would be next to impossible to observe, in nature, a dozen different aquatic species in their natural habitat, posing perfectly together and all in focus at one time—but such a scene can easily come to life in an illustration. An illustration can clean up a messy dissection, portray muscles beneath the skin, or make the layers of an archaeological dig magically float in space. Even the instruction manual that comes with a new electronic device includes an illustration, because it's easier to see important features without the distractions of photographic shadows and detail. Because no one understands your research the way you do, it pays to think ahead about how you are going to picture your results.

TAKING NOTES IN THE FIELD

I make notes and drawings in the field—and for me "in the field" may simply mean in the presence of an actual, possibly live specimen—to understand and record information about my subjects' proportions, key features, colors, and even behavior. I recommend using materials that are lightweight, portable, and simple. The sketchbook I like best has a removable spiral binding, which allows me to carry drawing, tracing, watercolor, grid-lined, waterproof—virtually any combination of papers— all in a single book. I prefer to use a mechanical pencil when in the field, as it requires no sharpening. A black ballpoint pen is a wonderful tool for making shaded sketches: changes in pressure produce a surprising range of values from light to dark, and the ink doesn't smudge as easily as pencil. Finally, a permanent felt-tip pen with an extremely fine point and some type of color medium complete my basic sketch kit.

To start the process of familiarizing myself with a subject, I study reference materials before going out. A review of species descriptions and photographs, if available, helps me begin to form a mental picture of the organism. This process can also call my attention to gaps in information, alerting me to particular features to look for and questions to ask. For example, in photos of four-legged animals, feet are often obscured by grass. In the presence of the real animal, I would make sure to pay serious attention to the feet!

My studies of live animals usually begin with a series of quick draw-

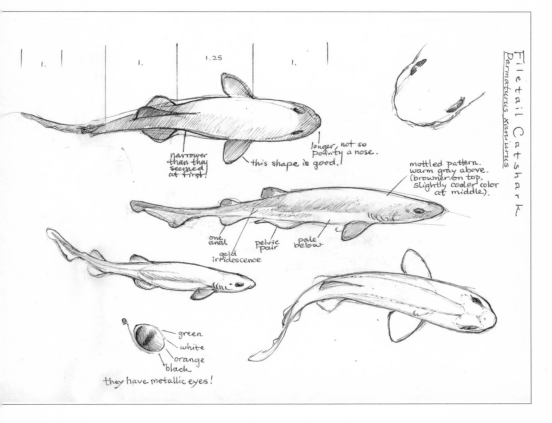

Sketches and notes on the filetail catshark (*Parmaturus xaniurus*) made with one of my favorite tools, a ballpoint pen.

ings, all on a single page. When the subject changes position—which it pretty much always does—I abandon the first sketch and start a new one. Continuing on in this way, the page fills with mostly unfinished squiggles until the animal eventually resumes one of its earlier poses. If that happens, I add as much detail or refinement to one of the earlier sketches as I can. While this is all going on, I also jot down written notes that help explain what I observed and what seemed significant. My visual notes tend to be peppered with arrows, diagrams, scale bars, corrections, question marks, and exclamation points. Back in the studio, these scrawled pages are among my most valuable and trusted resources in creating a finished illustration. They may not be complete or excruciatingly detailed, but when I review them I can really remember what I saw. My sketches—or perhaps more important, the deep observation that went into making them—also help me understand and interpret other visual references such as photographs.

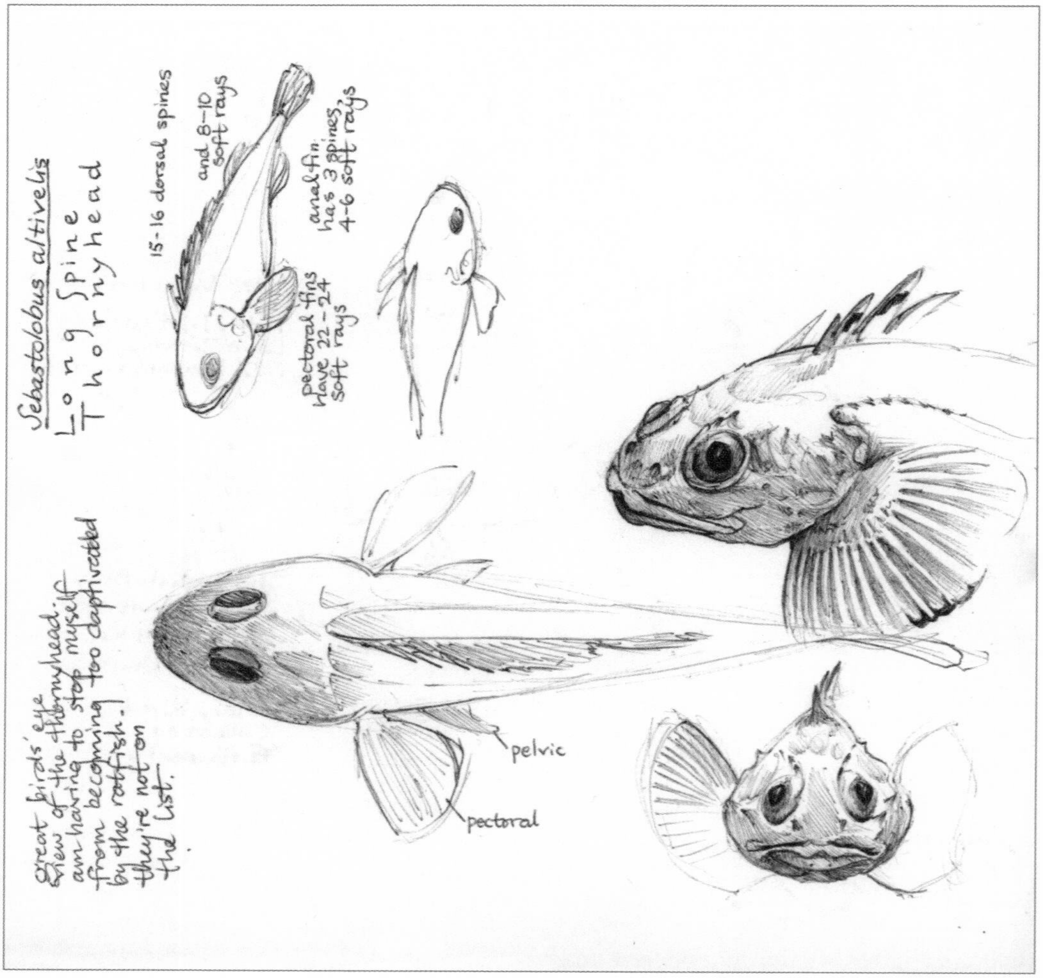
Sebastolobus altivelis
Long Spine Thornyhead
15-16 dorsal spines
and 8-10 soft rays
pectoral fins have 22-24 soft rays
anal fin: has 3 spines, 4-6 soft rays
great birds' eye view of the thornyhead. am having to stop myself from becoming too captivated by the rockfish. they're not on the list.
pelvic
pectoral

The range of values possible in a sketch made with a standard black ballpoint pen is evident in these sketches of the longspine thornyhead (*Sebastolobus altivelis*).

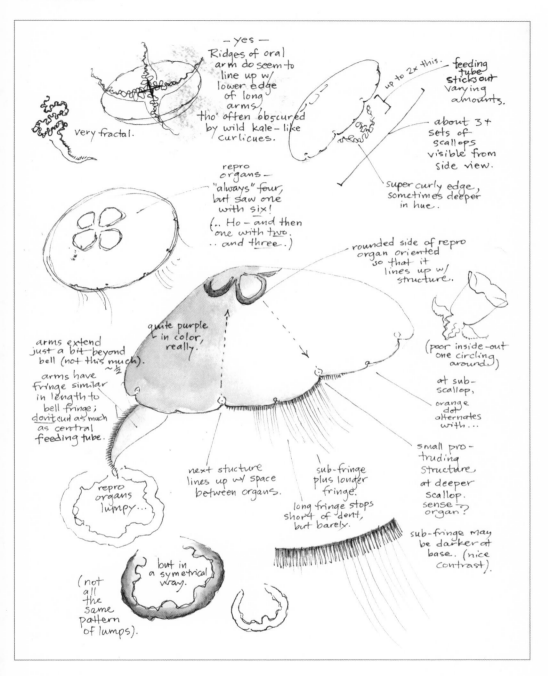

The written notes that accompany these moon jelly (*Aurelia* sp.) sketches supply additional
information, and also call attention to important details in the drawings.

After observing overall morphology and important features, I make notes about color. In-person, live observation of color is a practice for which I feel there is no adequate substitute. On a collected specimen, changes in color usually commence soon after an organism has died. This process accelerates—depending on the method used—when the creature is preserved, and inevitably continues as time passes. Photographs, as mentioned above, are often imprecise in reproducing color. Therefore, the live situation offers the best possible opportunity for seeing and describing hues correctly.

There are several ways to collect notes about color. Colored pencils are a good medium to start with: they are easy to use, portable, and come in lots of vibrant, natural hues that may already match your subject matter. Colored pencils can also be put down in layers, so you can modify one color with another. A word of warning, though: avoid getting a cheap set from the office-supply aisle of the drugstore. The results won't be worth the effort. Artist-quality brands have a higher concentration of pigment that will allow you to match what you see more easily, plus they can be purchased individually or in sets, making it possible to tailor them to your needs. Good colored pencils are worth the extra expense.

Personally, I usually use watercolors in my colored sketches, because a small paint set is even more portable—and faster to use—than colored pencils. Watercolor takes a little getting used to, but with practice one can mix exactly the color needed and then cover a large area with a sweep of the brush. With the watercolors I bring a "waterbrush," an ingeniously designed paintbrush that holds water in a hollow, plastic handle. Water is added directly to the dry cake of paint from the brush, transferred to the palette for mixing, and then applied to the paper. A scrap of cloth or a paper towel is used for "rinsing out" the bristles.

A third way to record color in the field—and one that requires no artistic training at all—is the use of a standardized, commercial color-matching system. The Pantone Color Guide, for example, is a loose-leaf book filled with hundreds of numbered sample colors printed on perforated pages. These pages can be held up to a specimen and when a color match is found, a "chip" of that color can be punched out and taped directly onto a page of one's notebook. A collection of such chips accompanied by written notes can very accurately describe the colors of

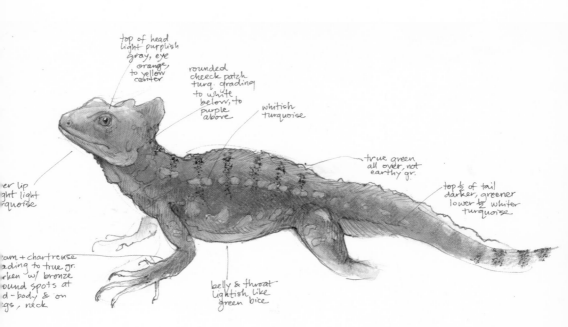

top of head
light purplish
gray, eye
orange,
to yellow
center

rounded
cheeck patch
turg. grading
to white
below; to
purple
above

whitish
turquoise

true green
all over, not
earthy gr.

er lip
ght light
rquoise

top £ of tail
darker, greener
lower to whiter
turquoise

am + chartreuse
ading to true gr.
rken w/ bronze
ound spots at
d-body & on
gs, neck

belly & throat
lightish, like
green bice

The wide range of
color saturation that
can be created with
high-quality colored
pencils is evident
in this drawing of a
basilisk lizard made
from a captive speci-
men at the California
Academy of Sciences.

a subject. In addition, these color descriptions can be precisely communicated, even across long distances, to anyone who has a copy of the same book.

A LITTLE DRAWING INSTRUCTION

Most of us have had the opportunity to watch an artist drawing from observation. He or she will look at an object, then down at a blank sheet of paper and, seemingly out of nothing, generate lines and shading on a flat surface that accurately represent a three-dimensional form. Especially if the drawing turns out well, this can seem like an almost magical skill. But what happens in the mind of an artist at work is not so mysterious. It's the same way you tackle any complex task: you break the problem down into smaller, more manageable pieces. In the case of drawing this means sizing up and actually *measuring* (either physically or by eye) the object in a variety of different ways to determine where each mark should go. These measuring techniques are something anyone can learn.

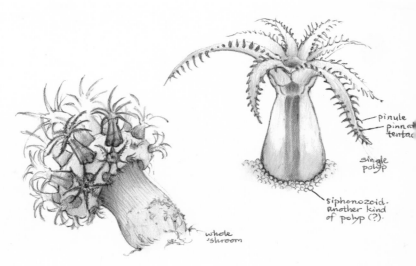

Mushroom coral and close-up view of polyp. The gradations of color in this animal were created by layering red, pink, and beige colored pencils.

pinule
pinnat
tentac

single
polyp

siphonozoid.
another kind
of polyp (?).

whole
'shroom

Mushroom Coral
Anthomastus ritteri

Lots of color variations:

mushroom cap	polyps
cream	coral pink
coral pink	white
reddish	reddish

Spotted Jelly!
*Mastigias
papua*

Watercolor painting of the spotted jelly (*Mastigias papua*). I often use watercolors to enhance quick sketches in the field because small paint sets and "waterbrushes" are even more portable and faster to use than pencils.

Kelp Greenling

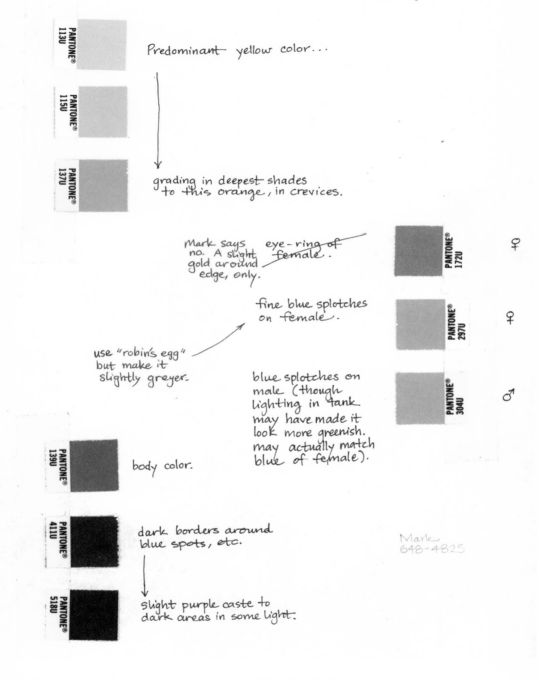

PANTONE® 113U

Predominant yellow color...

PANTONE® 115U

PANTONE® 137U

grading in deepest shades to this orange, in crevices.

Mark says no. A slight gold around edge, only.

eye-ring of female.

PANTONE® 172U ♀

fine blue splotches on female.

PANTONE® 297U ♀

use "robin's egg" but make it slightly greyer.

blue splotches on male (though lighting in tank may have made it look more greenish. may actually match blue of female).

PANTONE® 304U ♂

PANTONE® 139U

body color.

PANTONE® 411U

dark borders around blue spots, etc.

Mark 648-4825

PANTONE® 518U

slight purple caste to dark areas in some light.

These Pantone color chips and notes were matched to a live fish to determine the appropriate colors for an illustration of the kelp greenling (*Hexagrammos decagrammus*), a common fish on the coast of California.

Most artists look at the overall form of an object first, then work their way down to smaller and smaller details. Outlining the basic shape accurately is often the most difficult part of the job. If done correctly, smaller details should fit neatly into place like the pieces of a jigsaw puzzle. Below are a few visualization techniques that can help you see an object like an artist does when he or she is figuring out an accurate outline.

To draw this fossil brachiopod, first observe its general proportions. Hold the pencil up to the subject and use your thumb as a marker to compare its width to its height. Make a mental note of this ratio. On your paper, use that ratio in placing a few dots to indicate the general proportions. (Note that you don't necessarily have to draw the object at actual size—you only have to get the ratio right—the height in correct proportion to the width.) Make all these early marks on the paper very light so you won't have to erase them later.

Next, try to see the basic shape of the subject. The pencil can be used again as a visualization tool, helping to simplify curvy edges into straight lines and angles so a general geometric shape can be drawn. Draw these lines so that the basic shape fits within the dots already marked on the paper.

Use the pencil in yet another way to help see alignments—features that are directly across from each other, or directly above and below each other. There are no rules about which features to pick—any that are fixed in place and are easy to see are good. Mark a few dots here and there to indicate alignments on your drawing.

Now focus on the negative spaces surrounding the object, to help in refining the outline. What you want to do here is stare (really stare, and for more than just a moment) at the spaces around the object, until you can see them as shapes in their own right. Even try describing the shapes to yourself as you observe: "this one looks like a longish vertical triangle with a bite out of it." Once you have seen a negative-space shape clearly, draw its contours in the correct location on the paper. In doing so, of course, you will also be drawing the edge of the adjacent positive shape—the object—at the same time.

By the way, the reason to look at an object in a negative-space way is to give yourself a fresh perspective on the form. It's like drawing the iris of someone's eye (which you know is a circle) by drawing the shapes of the white areas around it. If you copy the white shapes accurately—and

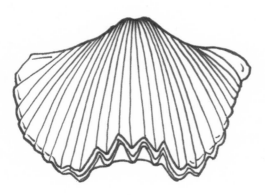

A fossil brachiopod specimen used as a model for drawing instruction.

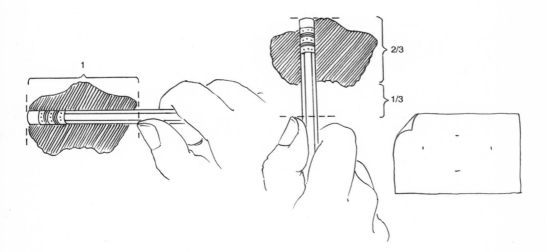

Step 1. Proportions. Determine the relative proportions of the specimen by using a pencil and your thumb to compare its height with its width. Make dots on your paper to estimate this ratio.

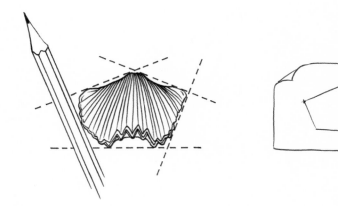

Step 2. Basic shape. Use the pencil to help visually simplify complicated contours, and imagine the specimen reduced to a basic shape. Transfer straight lines to the paper, fitting them within the dots made in Step 1.

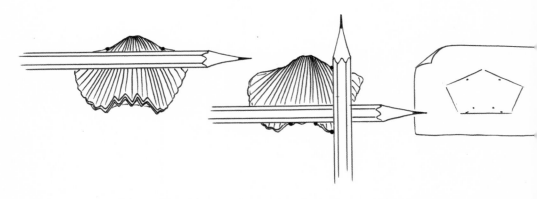

Step 3. Alignments. Use a pencil to determine where notable parts of the specimen align, and indicate them on the developing sketch.

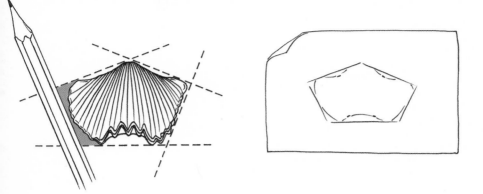

Step 4. Negative space. Stare at the spaces around the object to determine the particular shape and size of each. Incorporate these into the drawing.

place them in the correct orientation to each other—you'll discover that part of the iris-circle in a relaxed eye is cut off from our view by the upper lid. Looking at and drawing negative space can help us set aside what we think we know and pay attention to what we actually see. This is one of the foundations of good artistic (as well as good scientific) observation.

Finally, erase any extraneous marks and fine-tune the nuances of the outline.

Basic outline of the fossil brachiopod. Using the guidelines established thus far, focus on and draw the finer contours of the specimen. Interior details can be added by repeating the methods described above.

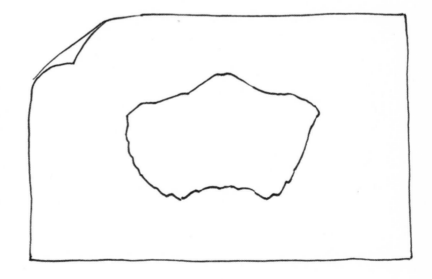

If you are wondering how one would go about finishing the interior details of the brachiopod, the method is essentially the same. Rather than attempting to draw all the radiating lines in one go (which would most likely result in a skewed perspective), break the task down into more manageable pieces. Draw, for example, the one line that divides the form in half. Then, draw the lines that divide those halves more or less into halves. Remember to stare at the negative spaces around each line to help you get its correct angle and curve—you may have to pretend there are no other lines except the one you are working on at that moment. Continue dividing spaces with lines until all the divisions are represented.

USING COLOR

Although the study of color mixing is too far-reaching to be adequately addressed here, a few words of advice can help you get started. Most important, value your intuition. If you perceive a color to be more of a greenish yellow than an orangish yellow, you're most likely right. Reach for the closest color in your pencil or paint set, then modify it as necessary.

Don't mix the three primary colors (that is, lemon yellow, magenta, and cyan blue) all together at once, unless you want a shade of brown. This sounds easy enough, but sometimes it gets tricky in execution. For example, we all know that red and blue mixed together make purple. However, if the red you choose has even a little bit of yellow in it (if it's what we would call a "fire-engine" red), your mix will contain all three primaries, and the result will be a dark brownish purple. To create a purple with true brilliance and clarity, mix the blue with something more like magenta, and keep all traces of yellow out of it.

Although it seems like a good idea, don't shade with black. If you want to darken or dull a color, add its complement—the color that is directly opposite it on the color wheel. It sounds crazy, but you can produce a natural-looking shadow on green by adding a touch of red or reddish brown, or effectively shade a yellow area by adding a very small amount of pale lavender.

Remember that colors in nature are, in general, more subdued than what comes directly from the pencil or the tube of paint. Greens,

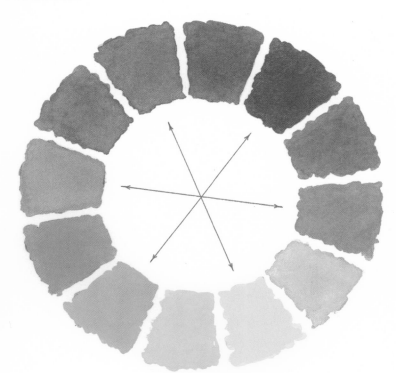

A basic color wheel. Arrows indicate complementary color combinations, which are useful for creating shadowed areas and subtle, natural-looking hues.

especially, tend to be a lot browner than we expect. Tone things down a bit when they need it.

DRAWING SHORTCUTS

I recommend thinking about field sketching as just another way to collect information rather than an effort to make pretty pictures. If the sketches turn out to be beautiful, that can be a bonus, but too much concentration on aesthetic goals usually just gets in the way of the job. When I release myself from caring about the prettiness of the results, I often work faster and better. I also feel free to take time-saving shortcuts with the drawing process.

The first such shortcut is to leave the drawing unfinished. Record as much information as you need, but don't draw any forms, details, or colors that are merely repetitive. The back and front of a representative flower on a plant, for example, or half of a bilaterally symmetrical animal may be all that's necessary. In some cases you

Since this sphinx moth is bilaterally symmetrical, the drawing does not need to be completed on both sides. This will save a good deal of time in the field.

may even forgo the drawing of the form itself. Rather than adding color to a drawing of an entire subject, make a set of color swatches that show the range of hues.

Keep it diagrammatic—that is, don't feel obligated to create a detailed, shaded representation when a simple line drawing will suffice. In many cases, basic structures and their relationships can be shown more quickly and clearly in a diagram.

Make use of references you already have before you go out. For example, trace an outline of your subject from a reasonably good photograph and, on the same page, write down questions you wish to investigate while looking at the real thing—perhaps about features that are unclear in the photos or in the existing literature. Take this annotated sketch with you into the field and use it as a base layer on which to record fresh observations.

Finally, consider the image-making potential of the subject itself. Mycologists will immediately think of spore prints, made by placing the cap of a mushroom over light and dark papers and leaving it to rain down its characteristic pattern of spores. While mushrooms are

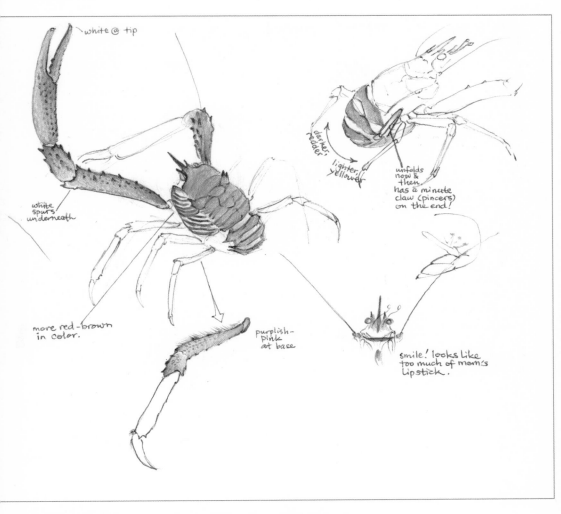

The handwritten notes on the image read:

white @ tip

white spurs underneath

more red-brown in color.

darker, redder

lighter, yellower

unfolds now & then, has a minute claw (pincers) on the end!

purplish-pink at base

smile! looks like too much of mom's lipstick.

Visual notes on a squat lobster (watercolor) demonstrate how drawings are often not completed: only aspects that illustrate necessary information are included.

unusually cooperative with this process, images can also be coaxed from less obliging subjects. A quick outline drawing of a leaf can be obtained by simply pressing the blade flat on the page and tracing around its margin. A perspective outline of a handheld object can be traced from a shadow cast by the sun onto white paper. Textures that may be difficult to photograph can be transferred to paper by laying a thin sheet over the object and rubbing the surface with a crayon. I have transferred to paper pale birch-bark patterns, incised marks on cylindrical pieces of pottery, and characters carved in the stone floor of a centuries-old darkened

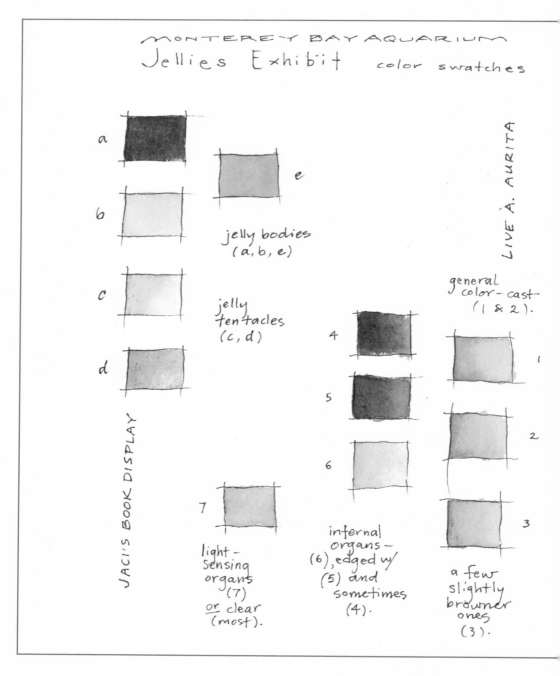

MONTEREY BAY AQUARIUM
Jellies Exhibit   color swatches

a

b

e

jelly bodies
(a, b, e)

c

jelly
tentacles
(c, d)

d

LIVE A. AURITA

general
color-cast
(1 & 2).

4

5

6

1

2

3

JACI'S BOOK DISPLAY

7

light-
sensing
organs
(7)
or clear
(most).

internal
organs –
(6), edged w/
(5) and
sometimes
(4).

a few
slightly
browner
ones
(3).

Simple color swatches can be used to document the range of hues
present in a subject—a type of visual shorthand.

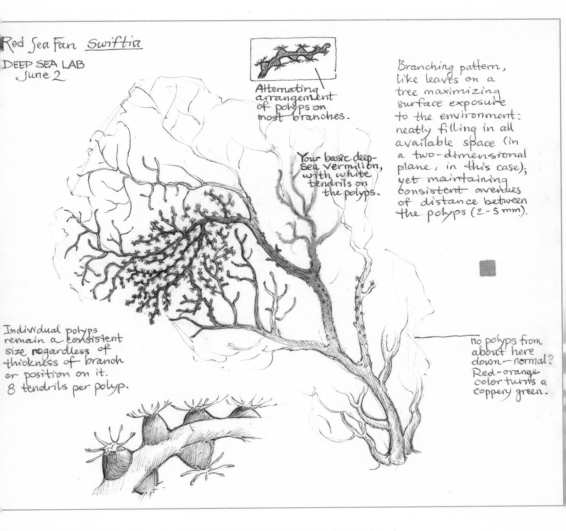

Red Sea Fan *Swiftia*
DEEP SEA LAB
June 2

Alternating arrangement of polyps on most branches.

Your basic deep-sea vermilion, with white tendrils on the polyps.

Branching pattern, like leaves on a tree maximizing surface exposure to the environment: neatly filling in all available space (in a two-dimensional plane, in this case); yet maintaining consistent avenues of distance between the polyps (2-5 mm).

Individual polyps remain a consistent size regardless of thickness of branch or position on it. 8 tendrils per polyp.

No polyps from about here down—normal? Red-orange color turns a coppery green.

This ink and watercolor drawing of a red sea fan (*Swiftia* sp.) represents the depiction of a complex organism with only selective detail.

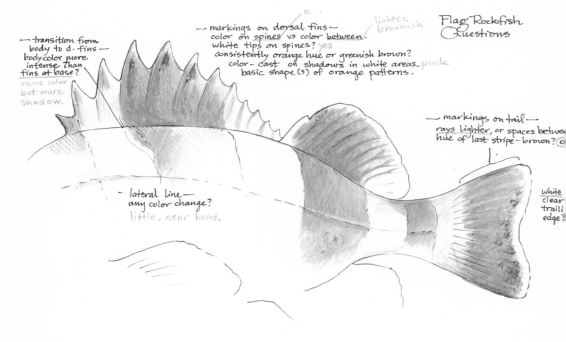

(above and opposite) **Notes on the flag rockfish (*Sebastes rebrivinctus*). The basic sketches of this fish were made from photographs before heading into the field. These were then used as a working "map" on which to record observations (in blue) from the live animal.**

church by making rubbings. You may feel a little silly as you get out your crayons, but there is actually less human interpretation and potential for error in a carefully made rubbing or tracing than in a sketch where proportions are measured purely by eye.

THINKING LIKE A SCIENTIST

The scientific thinkers quoted at the beginning of this chapter appreciated the value of drawing as a means of deepening their understanding of the natural world. At one time, drawing was considered an important, at times essential, part of the scientific process. Indeed, the history of science is filled with examples of drawn images that played a significant role in the discovery and presentation of new ideas. This is not a coincidence. The creation of an accurate drawing requires a systematic

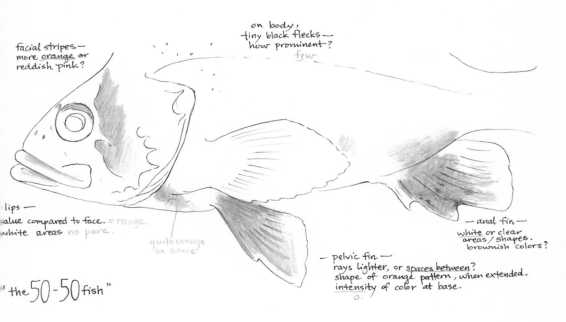

facial stripes —
more orange or
reddish pink?

on body,
tiny black flecks —
how prominent?
few

lips —
value compared to face. = range.
white areas no pure.

quite orange
on some

" the 50 - 50 fish "

— anal fin —
white or clear
areas / shapes.
brownish colors?

— pelvic fin —
rays lighter, or spaces between?
shape of orange pattern, when extended.
intensity of color at base.
o.

approach, patient observation, an openness to unforeseen possibilities, an ability to regard a topic from a variety of perspectives, a willingness to pay attention to both the exciting and the mundane, and the deliberate setting aside of preconceived ideas. Of course, all of these are useful approaches to the study of science as well. In my illustration courses over the years, I have witnessed scores of science majors discover with surprise and delight that they could learn to draw well, and that it had practical applications for them in their scientific careers.

9

The Evolution and Fate of Botanical Field Books

JAMES L. REVEAL

BOTANICAL FIELD BOOKS ARE A DEEPLY PERSONAL CREATION. There is no model, standard, or requirement for creating or maintaining notes in these books. That they are valuable and useful goes without saying—to future researchers and historians—but even now they are rapidly fading from the traditional norm of handwritten words on pages in a bound book.

My own effort began in 1958 as part of an advanced biology class at Sonora Union High School in California and continues today as I sit at a computer at Cornell University. My experiences probably are fairly typical of my generation of botanists. Field books were a required part of any class in plant taxonomy, and my high school teacher, Mary Long, felt it was a good practice. Where and when each specimen was gathered was recorded in a small pocket notebook. We were not required to create individual specimen labels in high school—we merely rewrote the infor-mation from our notebooks onto a sheet of typing paper to which our specimens were attached for her review. In those days, our guide to the California flora was Willis L. Jepson's *A High School Flora of California*, published in 1935, and it covered mainly the spring and early summer flora of the foothills and Sierra Nevada where I lived. I still have the collection—misidentifications and all!

At Utah State University, things became more formal under the tute-lage of Arthur H. Holmgren. Holmgren and my father knew each other from their days at Utah State, my father studying forestry and Arthur majoring in botany. My dad introduced me to Holmgren when I went to

Utah State in 1960. Holmgren was a large, strong man as a result of his years of playing football at the University of Utah, but he had a kind and gentle manner along with the gift of being a superb teacher. Little did I realize at that time that Arthur would ultimately see me through a master's degree at Utah State. Under Holmgren, one's field book included more than just a location; details were required, such as the ecology, elevation, and habit of the plant. Labels were then to be typed and submitted with the collection of fifty different species we made. In looking back at those pocket notebooks today, I realize that they were the simplest of entries, for I then believed I would remember all of the necessary information when I went to prepare the labels. Although these specimens went into the Intermountain Herbarium, I am not sure that my sparse notes enabled me to make good labels. I have since been afraid to go looking for them in fear of what I might find.

The end of my sophomore year was pivotal in my undergraduate career. I was taking taxonomy, a requirement for students in the forestry college then, and was finding a few curious items that I took to the herbarium for identification. While there I met Arthur Cronquist, who was in the West to collect plants for a proposed Intermountain Flora sponsored by the National Science Foundation. For some reason, Cronquist decided that I should become a botany major. In vintage Cronquist fashion, he went on and on and on until I gave up and agreed to change my major. Cronquist was already a taxonomist of considerable international fame at the New York Botanical Garden. He was a tall man with a loud, booming voice whose mere presence tended to dominate any setting. Later, I would learn that he was raised in Utah and attended Idaho State University with my father, who won a coin toss with Cronquist that allowed him to work on grasses, forcing Art to begin his long career on the sunflower family. If I had known more about my family history, I might have requested that a coin toss settle my somewhat one-sided discussion about my major with Cronquist. With the benefit of hindsight, I am glad that I didn't leave such an important decision up to chance. Having signed on to study botany with such a forceful endorsement, I was made aware of how critical were the proper methodologies of field books and active plant collecting.

On June 15, 1961, I made my first "professional" collection: "Number 189. *Polygala subspinosa.* In Rush Valley, Tooele Co., Utah, in the northwest corner of sect. 21, T.8S., R.3W." My notes indicate that the plant was found in loam soil in the middle of a dirt road associated with what is now *Achnatherum hymenoides—Oryzopsis hymenoides* in those days. I wrote down "about 5000" feet for the elevation; today, with computer-based maps, I see it was closer to 5,200 feet. My first specimen of *Eriogonum*, a genus that would occupy most of my subsequent years, was numbered "191" and was gathered that same day in the entrance to a still-unnamed canyon north of Bell Canyon in section 29. In September of 1961, Noel Holmgren (Arthur's second son) and I went collecting in southern Utah; my last number on that trip was 326, a collection of *Eriogonum cernuum*. It seems that I was just destined to work on that genus!

I maintained wire-bound pocket notebooks for my field data through much of 1964. Noel, then a graduate student with Cronquist, introduced me to serious note-keeping by using sewn Lietz field books, the same type of book used by surveyors for their work. My first entry, made on August 31, 1964, was a vast improvement.

Noel and I made a total of thirty-five individual collections from this location, referred to as herbarium sheets, as the plant represented a new variety of what eventually proved to be *Eriogonum brevicaule*. For some reason I did not record the township and range or the elevation in my field book—for the record they are T.7N., R.1W., sect. 5 and 9,400 feet.

The format of the field books was simple. On the right-facing page one finds the date, location data, associated species, elevation, comments about the plant itself, and its field identification. This information usually takes up several lines. On the left-facing page is the final identification and the number of specimens to be distributed to herbaria. From time to time I also added chromosome numbers and, when appropriate, indicated if the collection was a type specimen. If the final identification was made by another person, this too was noted.

The initial information, even today, is recorded in pocket notebooks. This includes simple notes about mileage, habit, and aspects of the plants, the local environment, and anything else that will help me to prepare specimen labels. Years ago, before GPS, laptop computers, and mapping software, elevations would be recorded from a handheld altimeter, and in the American West at least township, range, and section data were noted if there was a Forest Service or Bureau of Land Management

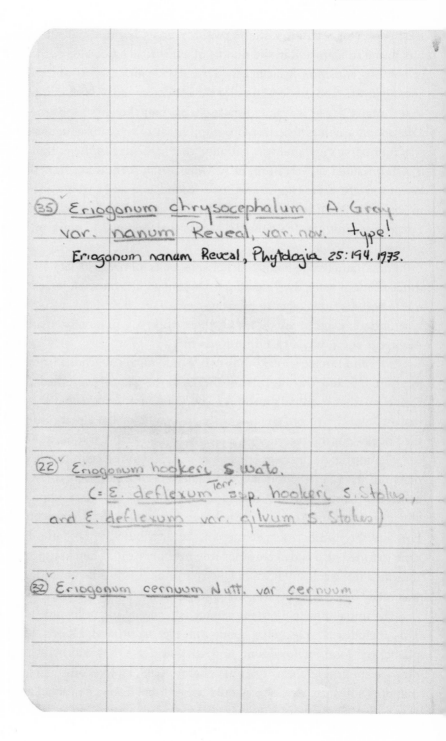

㉟ Eriogonum chrysocephalum A. Gray
var. nanum Reveal, var. nov. type!
　　Eriogonum nanum Reveal, Phytologia 25:194. 1973.

㉒ Eriogonum hookeri S. Wats.
　　(= E. deflexum ^Torr ssp. hookeri S. Stokes,
　and E. deflexum var. gilvum S. Stokes)

㉜ Eriogonum cernuum Nutt. var cernuum

Summer 1964 —
Collected nearly a thousand numbers with
Noel H. Holmgren.

Utah, Box Elder Co., On talus slopes
and marble outcrops south of Willard
Peak toward Ben Lomond Peak, on the
ridge top and adjacent slopes.
Aug. 31, 1964
665 *Eriogonum chrysocephalum* A. Gray
var. *nanum* Reveal, var. nov. (Type-resd Isotype)
   Associated with *Castilleja*, *Artemisia*,
and *Pinus*; common
   J. L. Reveal & Noel H. Holmgren
Forming mats 1-2 ft. across; calyx-segment
whitish-yellow; involucre 5-lobed.

Utah, Box Elder Co. Along Utah highway
70, 32 miles southwest of Rosette.
Aug. 24, 1964
666 *Eriogonum hookeri* S. Wats
   On sandy soil; infrequent

Utah, Rich Co. September 1st 1964
   1 mile east of Laketown
   667 *Eriogonum cernuum* Nutt. var. *cernuum*
      along roadside, associated w/ *Bromus*
   tectorum L., locally common

My first entry in a bound Leitz field book (August 31, 1964), made while collecting in Box Elder County, Utah, with Noel Holmgren.

map available with this information. The notebooks were also used to record photographs, travel expenses, and similar items that might eventually be useful. Looking back at these notebooks now, the information seems fairly sketchy, often abbreviated, and fairly uninformative. The purpose was merely to serve as a reminder for when, that evening, I would write up my notes in a proper field book. The field book itself was intended to be a permanent record so that anyone could understand what I found and where. I wrote it in pencil (preferably) or in ink, and tried to make sure the information was well organized and ready to be transcribed onto specimen labels.

When traveling outside the United States, I tended to use my field book as a journal, usually writing up the day's events after entering the collection data for that day. In looking back over these comments now, especially those made in Mexico and in China, I find sketches and helpful comments in spite of the misspellings and poor grammar, and occasionally I uncover some salient thoughts. Whether or not such comments will be useful to anyone in the future is dubious, but they are certainly useful for my own waning memory.

Journal writing was common in the past. In looking at important historical figures in the botanical exploration of the American West, such as Thomas Nuttall, David Douglas, and John C. Frémont, one finds that they often wrote true journals that were more like travelogues than records of what they had discovered.[1] Nonetheless, these journals provide useful information even when labels affixed to extant specimens bear little or none. These early collectors usually wrote in field books about the plants they saw without actually indicating that they made a collection at that place. This is understandable because most places in the West during the first half of the nineteenth century lacked geographic names. In looking at specimens in the Nuttall and Douglas collections, for example, their labels often will only say "Rocky Mountains" or "Northwest America." Thus, their extant journals are the only means by which one might find a more definite location. Douglas's journal is particularly good; unfortunately he lost a later journal when a canoe overturned, so one must rely upon letters that are not nearly as detailed. If Nuttall kept a journal during his travels in the West (1834–1836), it was lost or at least it has not yet been found. Fortunately, he was traveling with John Kirk Townsend, who wrote a book and kept a journal about his

154. Salvia clinopodioides Kunth

155. Salvia tiliifolia Vahl

156. Salvia polystachya Ort.

157. Stachys agraria Schlecht. & Cham.

158. Salvia leptostachys Benth.

14 October 1975
Mexico: Mexico:
Along Mexico Highway 55, about 14 miles south of the junction with Mexico Highway 15 in Toluca, on sandy-loam soil under pine overstorey.
4854. Salvia clinopodioides

The style and stigma is most interesting, being slightly curved and slightly apart horizontally, with downward curved hair, curving down one side of the upper lobe and the outer side of the lower lobe.

Along Mexico Highway 55 at Villa Guerrero, about 7 miles south of Tenancingo and 18 miles north of Ixtapan de la Sal, in weedy roadside habitat in a roadcut.
4155 Salvia
4156 Salvia
4157. Stachys

Along Mexico Highway 55 at La Calderon, 6 miles north Ixtapan de la Sal, on steep slopes above the river, associated with numerous shrubs.
4158. Salvia ~ Annual herb, spreading and rooting at the nodes.

Notes and sketches made on October 14, 1975, during a collecting trip in Mexico.

adventures from which one can get some idea where he and Nuttall were in 1834.[2] Frémont's published accounts are more detailed, and while it is often possible to precisely determine his route, he was not consistent about mentioning where and when he made his natural history collections. Although much of the available information on early botanists in the West (1790–1850) was summarized by S. D. McKelvey, information on later naturalists on the American frontier is more widely dispersed.[3]

Records made by early naturalists in eastern North America are even more scattered, and unlike their European counterparts, few made detailed notes in journals and letters.[4] At this time, there was little effort to maintain field books or to record one's natural history efforts in an organized manner. Still, in herbaria (especially The Natural History Museum in London) one can find a wealth of information on herbarium labels associated with specimens gathered from the 1680s up to the 1750s. In a real sense, the origin of journals as a place to record collections, and then field books, grew out of the need of European naturalists to have more detailed and precise information from the field collector all in one place.

Now that we are in the era of personal computers, traditional field books are being replaced by computer files. By default such "field books" are sterile creatures—all the words are spelled properly, the location data are exact to a matter of a few feet, and everything is properly formatted. In the spring of 1998, I penciled my last entry into my signature field book with the bright orange cover. Thereafter I have maintained a computer-based field book.

Oh, all the right stuff is there, clear, crisp and, well, dull. Every so often I print the pages so I have a hard-copy record. I even print the double-sided pages to fit the 4.75 × 7.25-inch page size of my field books, as if this makes it okay. There is still nothing more reliable than a printed page, given the propensity for computers and associated hardware and software to fail!

To compensate for having this new type of "field book," I have improved my note-taking in the field. Location information is written down in a notebook as a GPS entry—later I will compute distances (both road miles and straight-line air miles) using software. Township, range, and section data and other types of coordinates can be obtained from Graphical Locater.[5] Outside the United States, more exact information must be recorded in the field using available maps, odometer readings, and compass-based directions. All of these data can then be checked using Internet mapping sites. Also, digital cameras may be fitted with a GPS unit that will automatically enter GPS information onto an image file. This too helps with recording exact location data. Using a combination of your GPS data and Google Earth, one can actually pinpoint where a photograph was taken.

8482 *Astragalus mollissimus* Torr. var. *thompsonae* (S. Watson) Barneby (5)

Along Utah Highway 275, 1.1 miles east of the eastern boundary of Natural Bridges National Monument and 2.6 miles west-northwest of U.S. Highway 95, on sandy flats with *Juniperus* at 6725 feet elevation. N37°36′11″, W109°56′49″ - T37S, R18E, sec. 4 NW¼.

8483 *Astragalus coltonii* M. E. Jones var. *moabensis* M. E. Jones detr. S. L. Welsh (5)

Comb Ridge east of Butler Wash, along Utah Highway 95, 1.7 miles east of the Comb Wash Road and 12.4 miles west of U.S. Highway 191 at White Mesa, on sandy slopes among sandstone outcrops at 5200 feet elevation. N37°29′42″, W109°38′27″ - T38S, R21E, sec. 7 SE¼ of the NE¼.

8484 *Astragalus cottamii* S. L. Welsh (4) detr. S. L. Welsh

8485 *Phemeranthus brevifolius* (Torr.) Hershk. (1)

13 May 2004 (with C. Rose Broome)
COLORADO, San Miguel Co.:

Big Gypsum Valley, along Colorado Highway 141 at milepost 37, 7.2 miles southwest of Basin, above Big Gypsum Creek on low gypsum hills north of the road and just east of Road 23R, with *Atriplex* at 6400 feet elevation. N38°01′32″, W108°38′58″ - T44N, R16W, sec. 32 SE¼.

8486 *Cryptantha gypsophila* Reveal & C.R. Broome (27)

Big Gypsum Valley, along the S22 Road, 0.4 mile north of Colorado Highway 141, this junction 8.2 miles southwest of Basin, above Big Gypsum Creek on low gypsum hills east of the dirt road, with *Atriplex* and *Eriogonum* at 6300 feet elevation. N38°02′21″, W108°39′54″ - T44N, R16W, sec. 29 SW¼.

8487 *Euphorbia* (2)

8488 *Cryptantha gypsophila* Reveal & C.R. Broome (40) – Type collection

Big Gypsum Valley, along the S22 Road, 1.6 mile northwest of Colorado Highway 141, this junction 8.2 miles southwest of Basin, on low gypsum hills, with adjacent *Atriplex* and *Eriogonum* at 6270 feet elevation. N38°03′00″, W108°40′38″ - T44N, R16W, sec. 30 NENW¼.

8489 *Cryptantha gypsophila* Reveal & C.R. Broome (5)

8490 *Astragalus* (7)

Dry Creek Basin, along the 31U Road, 4 miles south of the U29 Road, 5 air miles southeast of Basin, on sandy soil at 7100 feet elevation. N38°00′48″, W108°28′37″ - T43N, R15W, sec. 1 SW¼.

An example of a computer-based field note page from a collecting trip in San Miguel County, Colorado, in May 2004. I transitioned from traditional field notebooks to computer-based field notes in 1998.

My digital field book is then generated from the actual label data, so that in my case I just never seem to write down any general observations or feelings like I once did. Whereas before I would have written a description of what I thought might be a new species in my field book, now it goes into a laptop. In this way I can still to California Academy, where John Thomas brought it down to Stanford where I was, reviewing that collection. I called Mary that same day, Saturday, 26 July, and arranged to collect it today, Tuesday. It is a new genus, and Mary is greatly pleased to have it named for her.

31 July 1975 — with John H. Thomas
CALIFORNIA: SAN MATEO Co.:
Jasper Ridge Biological Experimental Area, about 5 miles southwest of Palo Alto, on oak covered hillside south of San Francisquito Creek, on low, serpentine ridge at about 600' elevation. 3911. Eriogonum luteolum
Based on the observations made here, just outside Stanford University, it is clear that these plants are E. luteolum. The population has the two color phases of the flowers, but has narrower leaves

A field sketch of the inflorescence of *Dedeckera eurekensis* below my notes about Mary DeDecker's reaction to having the new genus named for her.

make measurements from hundreds of individual plants in the field, but I tend to be overly particular about it—the format has to be right, everything properly spelled, the descriptive sequence in the proper order, and even the observations drafted with the final publication in mind (rather than what I happen to see at the moment). The emotions of finding something new, once mentioned in my handwritten field books, are now missing, as if my mental editor says "no, that is not proper for a scientific journal."

WRITING SPECIMEN LABELS

Computerized production of specimen labels has improved greatly over the past decade and no doubt more technological advances will improve this process in the future. As more and more images of specimens are made available online, such data, when coupled with field photographs, will only result in more and better information for specimen identification. At Cornell University, the electronic Tompkins County flora allows one to pinpoint any specimen on an electronic map.[6] Thus, future photographs of that species and its population, taken when the specimen was gathered, will be available.

When I taught an undergraduate course in plant taxonomy at the University of Maryland, I provided students with a listing of required and optional items to go on a plant label. Those instructions, brought up to date, are still useful. Each label must contain, at minimum, the regional or national location where the plant was collected, its scientific name and the family name, ecological and habitat data, information about the plant, collectors and collection number, the date the collection was made, and an indication of the institution of origin for the specimen. This information should be formatted as follows: At the top of the printed label should be the name of the politically defined unit in which the plant was collected, and at the bottom of the label the institution or institutions of the collector. The family name should be printed on the line below the header, and the scientific name on the next line, centered and in italics or underlined. If there is an infraspecific name, it will be on the next line and centered.

In the body of the label, indicate in capital letters the county where the plant was found followed by a colon. Then give the location where the plant was found so that the spot might be found by someone else, and a person with a generalized road map could find the location again. Avoid such references as "hill near my house" or "front of grandfather's barn." Someone in the distant future may well be trying to find this spot again. Give the township/range, degree-minute-second, and/or the UTM (Universal Transverse Mercator) data. Add the elevation data either as a range where the plant is found or at the place where the collection was made.

Try to indicate where, at the locality, the plant can be found. This can often be accomplished by noting the type of soil or rock outcrop, exposure, or general condition where the plant is found. You can amplify this by indicating the associated plants. Likewise, a general statement about the abundance of the plant is helpful, especially to those in the future studying plant distributions.

Next, provide descriptive information about the plant that either is not obvious or might be lost in the future. Some taxonomists add local common names if they learn them. Observations are often useful and can help future scientists better understand the plant.

If the collection is used as a voucher, its designation should be reported stating what was gathered and for whom. Also, all persons involved with the collecting of specimens should be recorded (within reason). Most often the names of two to five additional collectors are recorded on labels. However, the primary collector's name appears first, followed by his or her collection number. All collectors assign sequential collection numbers to their specimens and never repeat numbers during their lifetimes. Some collectors have elaborate numbers like 20080001 for the first plant and the first collecting site in a given year. A large number may seem impressive, but simplicity is best here. Start with "one" and go from there.

The date the collection was made must be recorded on the label. Avoid using the form 08/09/08, for in the United States this usually means August 9, 2008, but in Europe this would mean September 8,

---

PLANTS OF NEVADA

7

Polygonaceae

Eriogonum contiguum (Reveal) Reveal

Nye Co.: Ash Meadows, along Ash Meadows Road toward Point of the Rock Springs and the headquarters of the Ash Meadows National Wildlife Refuge north of the Bob Rudd Memorial Highway, south of Bell Vista Road on alkaline flats associated with *Atriplex*, at 2225 ft elev. 36°23′ 12″, 116°18′06″, NW ¼ of sec. 13, T.18S., R.50E. DNA voucher for Elizabeth Kemptom (RSA).

James L. Reveal 8875   5 May 2008

L. H. Bailey Hortorium (BH), New York Botanical Garden (NY)
University of Maryland (MARY)

Example of a printed label for a plant specimen.

2008. The taxonomic community has largely adopted the mode of giving the day of the month first, followed by month (first three letters if abbreviated and then without a period), and finally the full year (2008, not 08).

While plant specimens with their associated labels end up in herbaria around the world, the fate of original field books often is less certain.[7] Some pass on to members of the family where their value may or may not be fully understood. Some end up in libraries at institutions with herbaria, or in the archives of herbaria. A few become part of a botanist's archive filed at that individual's home (or last) institution. More and more field books are going to major botanical repositories of historical objects, such as the Hunt Institute for Botanical Documentation at Carnegie-Mellon University, where they are curated and cataloged. At least in such permanent archives the existence of a person's field books becomes public knowledge.

The final resting place of field books obviously is the choice of the botanist, but first and foremost all naturalists should recognize that such books are vital historical records capable of providing significant information to future researchers. Aside from the need to sometimes compare information on a label with the original field book to resolve a seeming conflict, field books are often a source of information of interest to ecologists and biogeographers who are attempting to understand environmental changes in a given area. Having a listing of the plants collected in a single place, along with information about the associated species, can more easily be assembled from a field book than it can from searching herbaria for specimens from a single location. Also, field books with journallike commentaries can be of significance to historians and the occasional biographer.

How naturalists maintain field books in our modern era still remains a personal choice. In looking over my own forty-five years of keeping a record of plant specimens, I find that I am personally moving further and further away from the words I generate, becoming more aloof and separate from the experience of the actual event of collecting, concentrating instead on the precision of where and when. It is merely record keeping for the sole purpose of giving the facts.

I suppose I am a creature of my generation—raised with pen and paper rather than a keyboard and computer screen—and this colors my perspective. Maybe those who follow will not be intimidated by the computer, if that is the right way to say it, and their personalities and emotions will show through in the digital age. The problem with a modern computer-based field book is that it lacks anything personal about the writer (and doesn't encourage or accommodate other notes, such as, "Is this a new species? Carpels are similar to ————. Why might this be the case?"). With the decline of letter writing and the sterilization of field books, what we are losing is the individual. Field books are like letters that are replaced by often ephemeral emails. I fear that as we move further into the computer age we will similarly lose the detailed historical record that field books once provided. Sadly, the personalities of botanists will also be lost, for such musings as might be found in a field book are often telling to those wishing to know more of the past.

10

Note-Taking for Pencilophobes

PIOTR NASKRECKI

WHILE I CANNOT CLAIM that I invented it, I am pretty sure that I have perfected the horizontal filing system for papers, books, or any other objects that are flat enough so that they can be placed upon one another without falling, at least until a certain critical height is reached. This system, also known as the stratigraphic filing method, is based on the simple principle, analogous to sedimentation and familiar to every geologist, that the oldest documents accumulate at the bottom of a pile, with the layers getting progressively younger as they are deposited at the top. Of course, like in geology, violent movements of document strata often result in the disappearance of the youngest layers under a pile of overdue requests for manuscript reviews, old calendars (you never know when you may need them), and copies of a variety of papers, most of which can also be found online.

Having such highly evolved organizational skills, it is no surprise that for the longest time my note-taking in the field followed principles of progressively increasing entropy. This meant that I put my observations, measurements, or any other data on paper in a random, rather haphazard way on any available scrap of paper, and not necessarily in the order the observations were made. The system worked, sort of, but occasionally the sheet of paper on which I scribbled the temperature and codes for my sound recordings would never reappear after I returned from the field. I would, of course, remember where I kept it in my tent, but these documents that did not survive the trip home made the hours of stalking singing insects an exercise in wasted time. Even worse, while working on a faunistic survey of katydids and other orthopteroid insects, I would jot down the names and abundance of the observed animals only to lose the notebook, and with it days of observations. It was not rare for me to forget to write down geographic coordinates of a collecting locality, or put in the insect vials cryptic labels ("T17") and forget to write down the associated details about the origin of the specimens. Clearly, the situation was dire, and I needed help if I were to continue my work as a biologist, or as any other professional required to keep track of his own actions.

Luckily for me, two miraculous events coincided with the beginning of my doctoral program at the University of Connecticut, which, as seen from my current perspective, pretty much saved my career. One was the invention, and soon wide availability, of the portable computer, a laptop small enough for me to take into the field but expensive enough to force me to be always aware of its whereabouts. The other was the realization that I had the ability to visualize in my head, in a somewhat three-dimensional form, the relationships between and among disparate elements that make up a single observation event, which in turn allowed me to grasp instantly the basic principles of relational database design. In a rare flash of forethought I decided to eschew paper and from that point on to keep all my notes as digital entries on my laptop.

At that time, in the mid-1990s, there were not too many database managers available to biologists, but it just happened that my thesis adviser, Dr. Robert K. Colwell, aside from being a brilliant community ecologist, was also the developer of one of the first relational databases for biologists, Biota. Unfortunately, in its early stages of implementation, Biota did not yet have all the elements that my work on taxonomy, systematics, and behavior of katydids required. I decided to develop my own solution, and Mantis was born.

There are two main differences between flat file data storage, such as in a spreadsheet, and a relational storage system. In a flat file spreadsheet, records are simple strings of entries (a row of numerical or text fields across a number of columns) that are independent of each other. These can be searched and formatted at will, but it is very difficult to create synthetic overviews of data. For example, in a spreadsheet of specimens it is impossible to display all that belong to species A (which may come from several localities) and, at the same time, all that come from locality X (which may belong to several species). In a relational model, entries are combined into logical groupings that share common attributes, and specimens can be grouped together based on their identity or origin; these groups can overlap or not, but it is extremely easy to get simultaneous overviews of subsets of data according to many independent attributes. Another difference is the lack of redundancy of data in a relational database—you don't have to retype the information such as the species name for each record; once entered in the database, you simply refer to it (link to it.) This both reduces the potential for human error during data entry and allows for easy, global changes to your dataset. For example, a change in the spelling of the species name will be automatically reflected in all specimens linked to it. Mantis had originally started as a set of heavily scripted flat files that mimicked the behavior of a relational database, but it soon evolved into a fully relational database manager with a minimum of redundancy.

Having a single, centralized system for all my behavioral observations, taxonomic information, references, measurements, photographs, and sound recordings truly changed my life. Nearly everything I know about katydids, every single specimen I have ever looked at, coordinates for each site I visited, and every temperature measurement I have ever taken, all resides in my database. I can access them at any moment and take them with me everywhere I go. Mantis has become an extension of my brain, an extra memory storage space that never forgets anything and thus, I am convinced, is a reason for major memory lapses on my part. Why should I make an effort to remember the author of that paper on the courtship behavior of *Cyphoderris* when I can quickly look it up? But there is no going back now, and everything I do as a taxonomist and a field biologist revolves around and eventually ends up on the infinitely expandable virtual shelves of my database.

Before I get into the details of my note-taking process, a disclaimer is needed. What follows is a description of my use of Mantis, a database manager that I developed and make freely available to anybody interested in it, but it is not a sales pitch. There are many excellent database managers available to biologists these days, and if you would like to follow a similar route for your note management, I strongly urge you to explore all available options.

I am a taxonomist and a conservation biologist, working primarily on sound-producing insects with occasional forays into arachnology. Thus, the data I need and collect are related to taxonomic nomenclature, species distribution and abundance, behavior, host-guest relationships, and environmental threat assessment. Nowadays, most of my fieldwork consists of rapid biological surveys, conducted on behalf of either a conservation organization or mining companies, with the goal of creating a baseline biodiversity assessment for a series of biologically unexplored sites. The underlying objective of such surveys is always to maximize the number of recorded species and collect as much information as possible about threats to their habitat. At the same time, my interests in insect behavior and systematics require collecting additional pieces of information.

Therefore, the types of data I need to record in the field can be summarized as follows: geographic coordinates of each collecting site, site description (including a list of dominant plant species), evidence and type of human impact, species identification, species abundance (every observed specimen gets its own record), sex/stage of each specimen, date and time of species activity, singing activity data (duration and timing of the call, ambient temperature, type of the call, technical data of the recording), host plant data (for herbivorous species), and specimen collection data (collecting method, type of specimen preservation, unique specimen ID, and so on). This is a fairly long list of data points, but having a database designed specifically to record them simplifies the record-keeping process tremendously.

The first thing I do upon arriving at a new site is to record the GPS coordinates of the base camp and describe my first impressions of its surrounding vegetation (in the Locality and Event tables of the database, respectively). I do this immediately, literally minutes after pitching my tent. The habitat and vegetation descriptions are refined with time as I get additional insight from other participants of the survey, especially the botanists. Since each new entry into the database is automatically time-stamped, if need be, it is easy to reconstruct the time line of events.

I should mention that our team usually has a generator, which makes recharging the laptop battery possible. On those few occasions when we did not have one, I used a small, portable solar panel that produced enough power to recharge the battery every day. During the course of the survey, I establish a routine of updating my database records. New coor-

Specimen of **Hemihetrodes bachmanni (Karsch, 1887)** ☒

| pecimen card | DNA data | Images | Sounds | Associations | Species | Event | | Find flagged | Flag record |

① **②**

SPECIMEN DATA   Created 27 Oct.08   Modified 11 Feb.10

Record type ● **Specimen**   Stage | male   Recording | SAFRI-21
           ○ **Specimen lot**
                           Medium | song recorded   Depository | MCZ

③   Method | night collecting   Storage | SA-04

                           Type |

                           Det. by | P. Naskrecki   Custom 1 ○ Yes ○ No

IDENTIFICATION   EVENT DATA

Species | **Hemihetrodes bachmanni (Karsch, 1887)**   Country | **Republic of South Africa**

Senior | Province | Northern Cape   Elevation | 243

Tribe | Enyaliopsini   Locality | Richtersveld Heritage Area, 4.85 km SE of

Subfamily | Hetrodinae   Date | 1.x.2008   Stop | RSA_24A

④   Species ID | 21572   Collector | P. Naskrecki & C.   Event ID | 528⁷ **⑤**

| Link species | Edit species | | Link event | Edit event |

BODY SIZE   OTHER DATA

1. body w/wings | Notes | individual singing from a low branch of Acacia sp.; female about 1
2. ~~w/o wings~~ | 35   m away
3. ~~um~~ | 17
⑥ 4. ~~men~~ | 13   **⑦** Sp____
5. hind femur | 22.3   history
6. hind wing | Citation |
7. front femur | 12
8. mid femur | 12.2 | Link citation | Edit citation | Citation ID |
9. measurement 9 |
10. measurement10 | Extraction |
11. measurement11 | Recording |
12. ovipositor |
ratio [12] to [5] | 0.00   Custom 4 |

                                     Specimen ID | 14010004

| New specimen | Specimen labels | 45/51 | ⏮ ◀ ▤ ▶ ⏭ | Delete specimen | Find specimens | **MENU** |

"Mantis" data records for a specimen of the katydid *Hemihetrodes bachmanni* (Karsch) collected in Richtersveld National Park, South Africa: (1) images of the specimens; (2) sound recordings of the specimen; (3) basic specimen data and storage information; (4) identification of the specimen (linked to the taxonomy table); (5) description of the collecting/observation event and location data (linked to the event/locality tables); (6) physical measurements of the specimen; (7) additional notes about the specimen. Additional data (molecular sequences, host/guest associations with other organisms, links to published papers) may be added later; changes to the status of the specimen are automatically time/date stamped and recorded in the Specimen History field. See <http://insects.oeb.harvard.edu/mantis>.

Images: 2 ◄◄ ▭ ►► ⊖ ⊕ | Compare specimens | Specimen ID PNC001401 Hemihetrodes bachmanni (Karsch, 1887)

Image ID 179414   Specimen ID PNC001401
Category male
Caption habitus, dorso-lateral view
Type   Stage male
Source South Africa, Richtersveld
Copyright
Notes
Photo by P. Naskrecki
Created 10/12/2008   Modified 10/16/2008

Delete image | Add image | Specimen card | Find specimen

habitus, dorso-lateral view, male (Republic of South Africa: Richtersveld Nat. P., nr. Helskloof Gate, 29.ix.2008) (Source: South Africa, Richtersveld)

Photos associated with the specimen record of *H. bachmanni.*

dinates collected with a GPS unit are downloaded to my computer and the database as soon as I return from the field to the camp. Collected specimens (or at least one representative of each collected species) are photographed, assigned a unique specimen ID number (which goes into the alcohol vial or on the insect pin), and identified. Identification is often preliminary, and I create a temporary ID for each morphospecies. In many cases, however, I am able to identify species right in the field thanks to nearly 40,000 photographs of katydid type specimens and their diagnostic characters I have gathered in the database. I also carry with me PDF versions of important taxonomic papers that may help me identify specimens in the field. In the case of common, easily identified species that do not require physical collection, or if I am able to detect a species' presence by its song alone, I create observation records, equivalent to specimen records. Each specimen or observation record contains information about the specimen's ID, its sex/stage, when and how it was collected, its host organism (in most cases, a plant species, but sometimes a termite colony), the preservation method, and where it is stored. (I number each vial and box used to store specimens.)

The bulk of my note-taking activities take place during the day, when most katydids are hiding, and so going into the field to collect them is counterproductive. Thus, after all specimens have been prepped and photographed I usually have plenty of time to enter them into the database. Katydids are rarely very abundant, and my average workload is thirty to fifty specimens per day. Once this is finished, I transcribe the voice notes I took on a digital recorder the previous night while I was recording courtship calls of insects. Each recording is downloaded to the computer and linked to a record that includes such information as the identity of the caller, presence or absence of other individuals in the vicinity of the caller, air temperature, the distance of the microphone from the caller, technical data of the equipment used to make the recording, and time and date of the recording.

A sound recording and associated data of *H. bachmanni*.

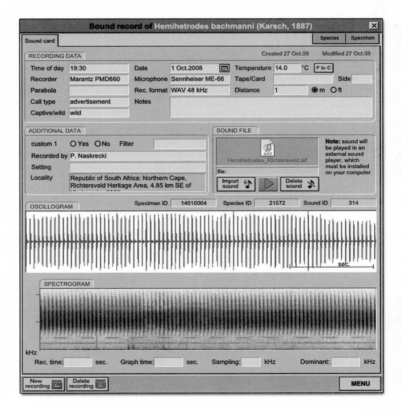

Any additional observations and data are also entered into the database as soon as possible. Of course, I do not carry my laptop with me when out in the forest at night, and if anything requires me to make a note I either record it as a voice message on the sound recorder (which I always carry with me), or make a note in a small, waterproof notebook. Over the years I have

A record showing the taxonomic classification of *H. bachmanni*.

trained myself to be obsessively conscientious about transcribing my notes, and as of yet, I have not lost any of my observations.

In the example illustrated, I collected a specimen of *Hemihetrodes bachmanni* (Karsch) in the Richtersveld National Park during a survey of katydids of South African Namaqualand. We went out collecting on a night in early October, and I used my acoustic recorder first to record its stridulation and then caught the insect and took a photograph. The next day, I uploaded the acoustic files and digital photos into Mantis, and recorded the specimen information.

At the end of the survey, all of my observations and identifications were already in the database, and upon my return to the museum there was little that I needed to do in terms of data entry. What was left was to change the alcohol in the vials, print permanent locality labels to replace the field labels, replace handwritten specimen

A record of a collecting event showing the habitat.

IDs with numerical barcode labels, and continue the identification process of specimens that I started in the field. Any new species that required a taxonomic description were scored for their morphological character states using a dedicated, descriptive module of the Mantis database, and a publication-ready description could then be generated with a single click of a mouse.

But there is no such thing as the perfect system, and relying entirely on a digital medium obviously carries its own risks. Hard-drive failures, power surges, or stolen equipment are very familiar caveats of the digital age, and I have experienced them all many times over the years. But the same survival instinct that moved me to develop a strict, digital note-taking protocol helped me become very disciplined when it comes to making backups of my precious data. Regardless of whether I am in the field or back at home, any time I make a change to the database, no matter how trivial, I immediately make two additional copies of the entire dataset. At any given moment I always have three independent, identical copies stored on three separate physical media (usually two computer hard disks and a USB flash drive). In addition to that, every month I burn archival copies of the entire database (which has recently passed the 10GB mark) onto DVDs. I follow the same routine when in the field, and on the trip back from the field I carry two copies of my data in my carry-on luggage, and a third copy in a checked bag.

The fact that the amount of knowledge and data that once required a roomful of books and filing cabinets can now be stored on a memory stick the size of your thumb is both exciting and somewhat sad. Instant availability and portability of data make research in the field infinitely easier for scientists, but what is lost is the feeling of slow accumulation of knowledge and the physical evidence of one's scientific prestige—the extensive shelves of important-looking volumes and journals. As hard as it is to believe, even my horizontal filing system is slowly becoming a thing of the past. The piles are getting smaller and it has been months since the last major pile collapse. There is no denying it, the era of paper is fading fast, and I can easily imagine a time when students will be perplexed by the strange, primitive implement known as the pencil. As far as I am concerned, this time cannot come soon enough.

11

Letters to the Future

JOHN D. PERRINE AND JAMES L. PATTON

IT WAS DUSK in the mountains of California. As the midsummer sun slipped behind the rocky crags, the field team from the Museum of Vertebrate Zoology (MVZ) conducted one of the day's most important tasks. After a long day of checking traplines and preparing voucher specimens, their camp was quiet except for the gentle rasp of pens on paper as the team recorded the day's details into their field notebooks. Upon their return to Berkeley, their field notes would be collected, bound, and archived along with the specimens that now lay neatly pinned upon the drying boards around camp. Perhaps, decades in the future, researchers would scour these notes for details that now seemed trivial, such as the coyote seen during the morning trap check or the trio of gray jays that visited camp during dinner. Since these field notes would become an integral component of the museum's record of the ecological conditions at this site, the team was careful to follow the content and style guidelines set by Joseph Grinnell, the first MVZ director. Once the day's notes were complete, the team members drifted to their tents, perhaps to read by the glow of a headlamp or listen to music. It was the summer of 2006, and the team was part of a small army of researchers, students, and volunteers participating in the Grinnell Resurvey Project, an ambitious effort to repeat the vertebrate surveys originally conducted by Joseph Grinnell and his colleagues nearly a century before.

Since the spring of 2003 this scene has repeated itself dozens of times, from the foothills of the southern Sierra Nevada to the lava outcrops of the Modoc Plateau in northeastern California. Underlying these efforts are a few simple questions: Has the vertebrate community at these study sites changed over the past century? If so, does the pattern of change shed any light on the possible underlying mechanisms? Which species seem particularly sensitive to environmental changes, and which are more resilient? The MVZ's collection of voucher specimens has been an important part of this resurvey effort, but the extensive field notes taken during the original surveys have proven to be invaluable. As such, this project represents an excellent example of the scientific value of thoroughly recorded and carefully archived field notes.

THE GRINNELL RESURVEY PROJECT

At this point I wish to emphasize what I believe will ultimately prove to be the greatest purpose of our museum. This value will not, however, be realized until the lapse of many years, possibly a century, assuming that our material is safely preserved. And this is that the student of the future will have access to the original record of faunal conditions in California and the west, wherever we now work. He will know the proportional constituency of our faunae by species, the relative numbers of each species and the extent of the ranges of species as they exist to-day.[1]

—Joseph Grinnell, 1910

Joseph Grinnell was the first director of UC Berkeley's Museum of Vertebrate Zoology, which was founded in 1908. As evident in the quotation above, Grinnell had a vision for a museum that was more than just a collection of specimens. Even in the early 1900s, California was clearly a land in transition: the human population was expanding rapidly and human enterprises such as agriculture, mining, livestock production, predator control, and hunting for the commercial market had left their scars on the landscape and its wild inhabitants. One of Grinnell's primary intentions in establishing the MVZ was to document the distribution, abundance, and variation of vertebrate species and their habitats across the landscape, in large part to provide baseline data for future comparisons.

Grinnell realized that the collection of specimens met only part of this goal. His teams preserved thousands of skulls and study skins to document the local species and their variation, but the specimens alone left little record of the context in which these organisms had lived. Nor did the specimens give any indication of how these animals had acted in life, such as their microhabitat associations, nesting preferences, vocalizations, mating rituals, and other behaviors. Documenting the natural history of the species, especially in terms of their specific ecological contexts, greatly enhanced the value of the collection. As Grinnell wrote in 1910, "It is quite probable that the facts of distribution, life history, and economic status may finally prove to be of more far-reaching value than whatever information is obtainable exclusively from the specimens themselves."[2]

These facts required more space than was available on the small paper tag attached to each specimen. Grinnell therefore directed all the MVZ field researchers to record detailed field notes during their expeditions. The notes described the major habitat types at each study site, the dominant plant species and associations, and the microhabitats for each individual specimen, along with behaviors such as vocalizations, foraging patterns, and courtship displays. Grinnell was interested in the entire vertebrate community at the study site, and since it was impossible to prepare voucher specimens of every species encountered, the field notes also included lists of species observed, such as the birds seen incidentally around camp, along the traplines, or during timed walks along specific trails. Added to the field notebooks of many individuals were detailed drawings of animals and their nests or burrow systems; transliterations of animal vocalizations; recountings of conversations with local ranchers, trappers, and other inhabitants about the species present; diagrams of the vegetation communities on opposite sides of canyons or along lake shores; maps indicating campsites and delineating trails traveled; and photographs of organisms and their habitats.

The information in the field notes was an essential component of the assessment of each study site, and like the voucher specimens themselves, the field notes were carefully preserved. Care was taken to use high-quality bond paper stock of a standard page size and to write with permanent black ink (those of us who have been around for some time remember well the admonition to "always use Higgins Eternal Ink!").

3 mi. N.E. Coulterville, Mariposa Co. Cal.
El. 3200 ft.   June 9, 1915.

Santa about holes seem to have begun
working last night (warmer weather is
now coming on) but no Perodipus were
caught. Only one has been taken here so far
Trap lines at this station.

Trap lines in red
Transition in blue
U. Sonoran in yellow

N
↑

yellow pine
+ kit-kit-dizze jump crupes

0   1   2   3
miles

contour interval 100 ft.

yellow pine
+
manzanita
W.W. Vireos.
Cassin vireos
Cabanis woodpeckers
Purple finches

meadow
manzanita +
yellow pine

DUDLEY'S

SMITH CREEK

COULTERVILLE ROAD

SCHOOL →

MEADOW

3200

3100

X camp
McCARTHY'S.

LARGE
PASTURE
meadow larks
lark + sparrows

BEAN CREEK
Red winged blackbirds.

yellow pine
+
Black oak
(few chipmunks)   3100 ft

From U.S.
G.S. Sonora
quadrangle.

3 mi. N.E. Coulterville, Mariposa Co., Calif.
El. 3200 ft.   June 8, 1915

X   Storer - I took a picture of a mourning
dove's nest built on a steep clay bank
of a barranca right on the ground. U.S.
32, Exp 1/5 sec.   Dist 8 ft. very bright
light.

meadow (short grass)

tall weeds

bare clay

BARRANCA   Dove's nest   10 ft
             [2 eggs]

small stream

Juncus grass

DIAGRAM
OF
MOURNING DOVE'S
NEST ON CLAY
BANK.

Mammals have been scarcer here than at
any other place I have ever trapped. It seems
to be a little too high for any number of
individuals of "upper "sonoran" species
tho Perodipus, Ground squirrels, Peromyscus
truei, cottontail & jack rabbits and probably
Perognathus are represented. Grey squirrels
are not abundant & chipmunks almost absent.
Juncos and Crested jays are rarely seen,
thrashers come to the very edge of the yellow pines,
long tailed chats & lark sparrows enter the
yellow pine area a short distance [1-3 miles.]

Three pages from the field notes of Charles L. Camp illustrating the type of detail contained in the "Grinnell-era" notes. (*opposite*) **General map (based on the Coulterville 15 min USGS quadrangle available at that time) of area between Coulterville and Dudleys, in Mariposa Co., with position of traplines marked, elevational contours, life zones, and so on given.** (*above*) **Written details of animals seen and a drawing of a mourning dove (*Zenaida macroura*) nest found by Camp and Tracy Storer in a clay bank.** [*continued on next page*]

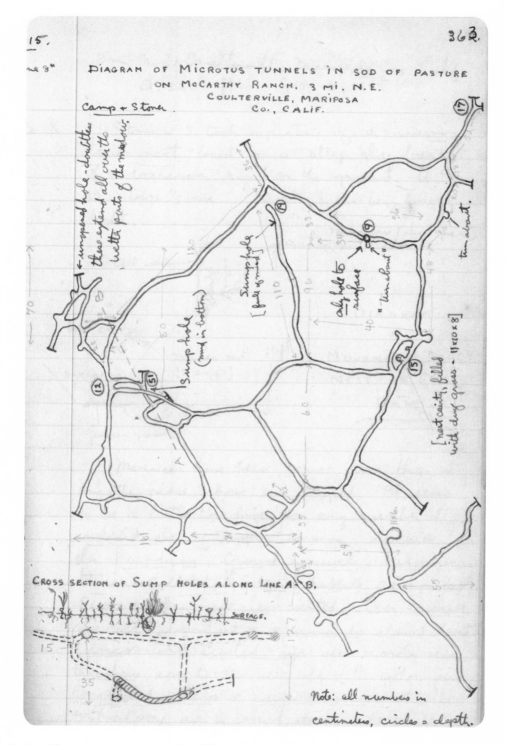

[continued from previous page] **Diagram of a California vole (*Microtus californicus*) runway system, horizontal and in partial profile to illustrate depth for that part belowground.** Archives of the Museum of Vertebrate Zoology, University of California, Berkeley, http://bscit.berkeley.edu/mvz/volumes.html; Charles L. Camp, 1914–1922, Section 3, pages 363–366.

To protect the delicate pages, Grinnell had them bound into hardcover books and archived alongside the specimens as a permanent part of the MVZ collection. The standard page size facilitated the binding process, and a standardized format allowed any subsequent readers to quickly find the specific information they needed. For example, Grinnell and other readers could use the date and locality information written at the top of each notebook page to easily link any specimen with its ecological context.

From 1908 until his untimely death in 1939, Grinnell and his MVZ colleagues traveled throughout California and other western states to document the distributions, ecological associations, and behaviors of hundreds of vertebrate species. From the accumulated collection of specimens and observations, Grinnell was able to make valuable contributions not merely in systematics but also in ecology, such as refining the concept of the ecological niche.[3] The fact that every vertebrate species had its own unique range and distribution contributed to the understanding that ecological communities were opportunistic assemblages of species rather than tightly coevolved "super-organisms," as had been earlier suggested.[4] The monographs describing the vertebrate communities of several focal regions, including the famous transects across the Yosemite and Lassen regions, as well as more local areas such as the San Bernardino and San Jacinto Mountains and the lower Colorado River, recounted numerous ecological and behavioral details recorded in the notes of the field teams.[5] Although some of these accounts are now a century old, for some of these regions and species the Grinnell-era monographs still represent the most comprehensive treatment of their vertebrate communities to date.

As Grinnell had intended, the depth and detail in these monographs, along with their accompanying voucher specimens and field notes, provided the comprehensive baseline for later comparisons. As the MVZ approached its centennial, a simple idea began to take shape: Perhaps it was time to return to some of these sites, repeat the surveys, and document what changes had occurred in the intervening decades. Although general scientific collecting had largely fallen out of favor in the middle of the twentieth century, the rise of the discipline of conservation biology and the emerging concern about the ecological impacts of global climate change made such a proposal more than mere whimsy.

The choice of where to start became clear when the U.S. Congress directed the individual National Parks to conduct an updated inventory of their biodiversity, which the parks are mandated to preserve for perpetuity. In Yosemite National Park, one of the jewels of the National Park System, the most comprehensive vertebrate inventory remained the 1924 monograph "Animal Life in the Yosemite" by Grinnell and his colleague Tracey Storer, which documented the results of their expeditions in 1914 and 1915. To the Yosemite Park biologists, a comparison of the current vertebrate fauna against the 1924 monograph seemed a logical strategy, and the ideal people to conduct the resurvey were the current researchers at the MVZ. Fieldwork began in 2003 and continued through 2005, and was by no means limited to Yosemite National Park itself. Instead, the MVZ team addressed the full range of the historic sites in Grinnell's "Yosemite Transect," which extended from the western foothills of the Sierra Nevada to the shores of Mono Lake near the Nevada border, encompassing nearly 1,550 square miles. The scope of the project entailed numerous challenges but provided a valuable perspective on the regional distributions of many species. A portion of the results, documenting that many of the small mammal species had experienced significant elevation shifts since the original surveys, was published in *Science* in 2008.[6]

On the heels of the Yosemite resurvey, the National Science Foundation funded the MVZ scientists to expand the resurvey efforts to other Grinnellian study sites such as the southern Sierra Nevada and the Lassen Peak region in northern California. This expanded effort has four primary goals: first, to document the current distribution of small vertebrates (principally birds and small mammals) at the historic sites; to compare the species' current distributions with their historic distributions documented by Grinnell and his colleagues; to link the observed faunal changes to changes in land use, climate, fire suppression, and other environmental factors; and finally, to provide a benchmark of current faunal conditions so additional comparisons can be made in the future.[7] As with the historic surveys, field notes continue to be the most essential component in documenting the ecological context of the species, whether preserved as voucher specimens or recorded only as observations. Such data enable future researchers to continue this process.

THE GRINNELL FIELD NOTE SYSTEM

Our field-records will be perhaps the most valuable of all our results; hence the importance of a convenient system of record.[8]

—*Joseph Grinnell, 1908*

The recording of field notes was common practice for biological surveyors and naturalists generations before Grinnell. The notes from Lewis and Clark's explorations of the Louisiana Territory and Charles Darwin's voyage on the *Beagle* are but two classic examples of the value of these handwritten records, and after nearly two centuries they remain engaging reading and relevant to science.[9]

The Grinnell field note system continues this tradition but is distinguished by its distinctive standardized format.[10] Standardization is essential in natural history collections, be they study skins, pinned insects, or herbarium sheets, especially when contributions are from multiple individuals. Otherwise, the resulting collection is a morass of different formats and styles, obscuring the very details that the collection is intended to reveal. By minimizing the distraction of varying styles, later researchers can efficiently see the patterns among the material regardless of who prepared the specimens. When researching his monographs, Grinnell likely spent many hours poring over his teams' field notes and cross-referencing the accompanying specimens, so he needed to be able to locate details quickly. In his role as teacher and mentor, Grinnell instilled the culture and process of field notes in a generation of ecologists and systematists, many of whom taught the "Grinnell system" to their own students.[11]

The Grinnell system field notebook consists of three sections. The *journal* contains a narrative account describing the study site and summarizing each day's activities and observations, including a list of species encountered. This section is often peppered with sketches, photographs, or maps. The *catalog* is a sequential record of all voucher specimens collected, each with a unique field number and the information needed for the specimen's museum tag, such as its sex, mass, breeding status, and standard body measurements. *Species accounts* are species-specific summaries of information and observations, gradually accumulated over multiple days at a site or across multiple sites, that eventually grow to detailed summaries of physical description, seasonal behaviors, microhabitat associations, and other characteristics. These accounts are a convenient way to organize the observations about a species, rather than having them scattered among the journal accounts for the various days, and to develop generalities based on these observations. The species account often represented the first draft of that species' treatment for an ensuing monograph. Regardless of the section, every page bears the author's name, the date, and the name of the study site.

Separating the notebook in this fashion allows each section to have its own context-specific structure and format for organizing qualitatively different types of information. The catalog has the most rigid layout, deriving from the fact that the same basic types of data are recorded in the same pattern for every specimen collected. Because of this, the catalog is the most analogous to a modern spreadsheet, with prescribed fields and formats for the date, location, catalog number, species, sex, breeding status, and several standard body measurements. Indeed, the information from the catalog is now uploaded directly to the MVZ's computerized specimen database. In contrast, the format of the journal is open-ended and flexible to accommodate whatever descriptive information needs to be included, such as written narratives, lists, sketches, maps, or photographs. These kinds of information are not easily transposed into the rigid system of datafields typical of a modern database. But like a database, the sections of the field notebook are linked by shared information, such as the name of a study site or the catalog number of a specimen. Using these links, the microhabitat associations for any specimen in the catalog can be cross-referenced to the journal narrative for that site and day, and to the behaviors and other accumulated information summarized in the corresponding account for that species. The usefulness of this system derives from the fact that the different sections and their associated formats allow for efficient data storage and retrieval by providing a balance of structure and flexibility. Other sources have covered the directions and mechanics of the Grinnell system and provide examples of the various sections and their respective formats.[12]

It is tempting to envision that Grinnell, with his keen attention to detail and deep insights into his institutional mission, dictated the details of the Grinnell system to his minions upon his first day as MVZ director in 1908. In reality, the format evolved over decades and generations. Grinnell himself never used the three-part journal system that now bears his name. He never wrote a species account, and his specimen catalog for each day was simply interspersed within his journal notes for that day, a practice he continued until he died. We can be sure that Grinnell gave some direction or instructions regarding the content and format of notes because virtually all of his colleagues and students from the early days of the MVZ used the same specific format and organizational structure. In every case, their field notebooks consisted of a simple narrative journal, with the specimen catalog interspersed within, and no species accounts.

One of the first to use the three-part system was E. Raymond Hall, one of Grinnell's strongest and most influential students. In 1928, the year he received his doctorate, Hall began both a separate specimen catalog and a set of species accounts. Interestingly, Hall did this only for his fieldwork in Nevada, which led to his treatise *The Mammals of Nevada*.[13] Hall's notes for other areas, even during the same time frame (1928–1941), were written just as Grinnell's: a simple narrative journal with specimen catalog interspersed within, and no species accounts.

By the 1930s, many MVZ researchers (with the notable exception of Grinnell himself) had adopted the three-part style of field notebook, with a separate specimen catalog, journal, and species accounts. The MVZ archives contain a memo by Grinnell, dated April 20, 1938, in which he outlined the standard procedures for taking field notes and for collecting and preserving specimens. Grinnell's guidelines did not mandate the use of the three-part system; in fact, the only mention of this system is in regard to the proper page numbering: "Number notebook pages consecutively from where left off previously. In notes arranged by sections, itinerary [the journal] precedes catalog, and species accounts are last."[14] Grinnell had his instructions printed on the correct size of paper so that everyone could include a copy in their notebooks to take to the field.

The three-part system became standard practice in the MVZ after Grinnell's death, as evidenced by Alden Miller's 1942 revision of Grinnell's instructions for note-taking. Miller, who succeeded Grinnell as MVZ director, gave specific examples of the appropriate page headings for the three sections, implying that it was no longer acceptable to integrate the catalog within the journal. Miller's support of species accounts

all around.   On the west-facing slope of the
White Mts. (Montgomery Pk. looming up. bare for its upper
third) one is able to see the timber belts
distinctly from a distance, as shown below.

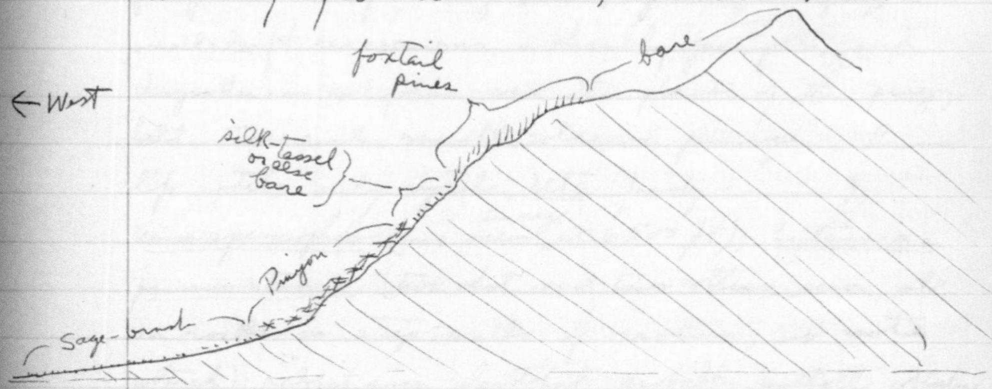

Dixon took a photo of this slope, and belts shown on
this should be compared with topographic map.
We saw no more signs of Citellus mollis than
around Pellisier Ranch.   Again we were told that
these animals were "all gone in" for the winter.
A very few birds were seen out on the sage flats:
Sage Sparrow (5 or 6), Brewer Sparrow (about 3),
Black-throated Sparrow (2).

                              Sept. 20
    A Kingfisher flew along the ditch at sunrise.
At daybreak heard Poorwills and Killdeer, and
saw 5 Mallards fly by.   Five magpies lit
silently on a dead cottonwood near camp, and
two Lewis Woodpeckers shortly took their places.
A Sparrowhawk percht, hunched-up, on the tip of
a dead tree; and meadowlarks began singing.

Woodhouse Jay (one along willow row, call very
like that of Calif. Jay); Magpie (3); Say Phoebe (2);
Parkman Wren (1); Warbling Vireo (1 in willow top);
Lazuli Bunting (1 seen clearly); Vesper Sparrow (4);
Intermediate Sparrow (about 6); Barn Swallow (2);
Savannah Sparrow (about 10).

✓ 4419, 4420 Uta stansburiana (2 specimens)

✓ 4421 Sceloporus biseriatus

✓ 4422 Calaveras Warbler ♂ im.   8.3g.

✓ 4423 Orange-crown Warbler ♂ im.   9.0g.

✓ 4424 Eutamias panamintinus ♂   45.5g.   188×82×31×12.5.

✓ 4425   "   "   ♂   47.3g.   190×84×30×12.

✓ 4426 Lewis Woodpecker ♀ im.   102.5g.

✓ 4427   "   "   ♂ im.   99.2g.

My trap-line got: big Perodipus 1♂; Onychomys
1♀ (put up by Dixon); Peromyscus sonoriensis 4 ad. ♀♀
(1 with 3 large embryos) and 3 ad. ♂♂. The Onychomys
was under an atriplex confertifolia bush, as was also the Perodipus.

✓ 4428 Perodipus (panamintinus?) ♂ 297×178×46.5×13.   60.6g.

✓ 4429   "   "   ♀

Sept. 21

✓ 4430 Perognathus panamintinus bangsi ♂ 7.7g.   133×72×19×4.

✓ 4431 Warbling Vireo ♀ im.   11.5g.   Shot by H.I. White yesterday.

✓ 4432 Western Robin ♀ im.   85.0g.   "   "   "   "

✓ 4433 Western Meadowlark ♂ ad.   116.0g.   "   "   "   "

My 30 trap-line produced: 1 Perognathus p. bangsi
(as above, under atriplex confertifolia), 1 Reithrodontomys ♂ ad. (on very
dry ground but where ditch from alfalfa had

Three pages from the field notes of Joseph Grinnell, written at Pellisier Ranch at the northern
end of the Owens Valley in 1917. (opposite page) Single page with a description of life zones on
the western flank of the White Mountains below Montgomery Peak. (above) Typical journal and
specimen catalog entries interspersed for the date of September 20, 1917. [continued on next page]

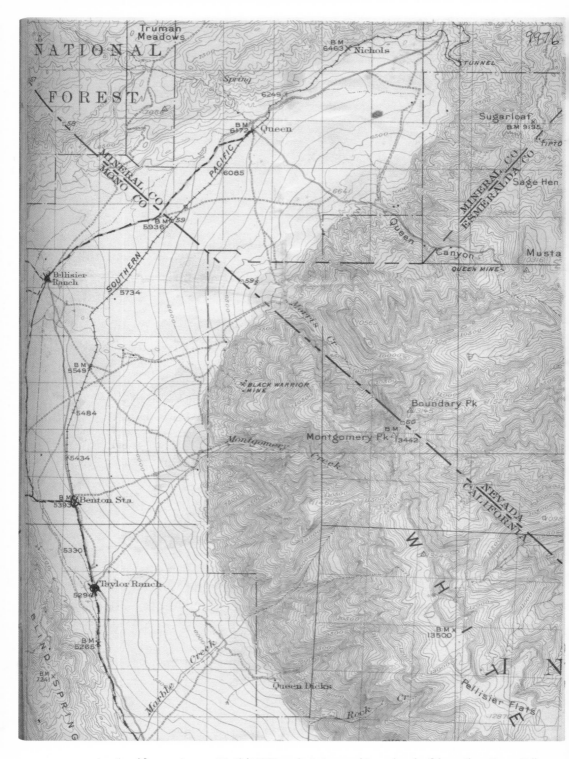

[*continued from previous page*] **Partial USGS 15 minute topographic quadrangle of the northern Owens Valley shows localities visited on this particular trip (Taylor Ranch, Benton Station, and Pellisier Ranch).** Archives of the Museum of Vertebrate Zoology, University of California, Berkeley, http://bscit.berkeley.edu/mvz/volumes.html; field notes of Joseph Grinnell, 1917–1918, Section 4, pages 1517, 1519, and map insert between pages 1521 and 1522.

was emphatic: "Enter as much information as is feasible under species accounts. It will be easier to look for it there at a later date" (underlining in original).[15]

Today, the three-part system has largely waned within the MVZ. Few MVZ researchers still write species accounts (neither of us ever have), but we all maintain a specimen catalog separate from a journal. Those who write species accounts tend to be studying particular taxa; thus, all of those observations would be in that taxon's species account, whereas the other details of where, when, how, and what else would be recorded in the journal. Although Grinnell established the basic guidelines and philosophy of taking field notes within the MVZ, it was his students (particularly Miller, Hall, and Seth Benson) who institutionalized the three-part system known today as the "Grinnell system." But despite the changes in format and organization over the years, Grinnell's original dictum of "record everything" remained the steadfast rule from then until today, and it certainly should continue to be the rule into the future.

FIELD NOTES IN THE GRINNELL RESURVEY PROJECT

> Remember that the value of our manuscripts increases as the years go by and faunal changes take place. Some of our earlier notebooks describe conditions now vanished in the localities they dealt with.
>
> —*Joseph Grinnell, 1938* [16]

It would be hard to overstate the importance of the historic field notes to the resurvey project. Indeed, the project could not be accomplished without the availability of these notes and the details contained within. We consult the field notes at virtually every step of the project, from the earliest planning stages to the preparation of the final reports. We bring photocopies of the notes with us in the field, where we review them almost daily for hidden details. Many institutions have extensive vertebrate collections, but few can match the depth of context for these specimens as provided by the written accounts in the field notes. In fact, given sufficient taxonomic expertise by the field teams, the resurvey could likely have been conducted without the original voucher specimens, but it would have been virtually impossible without their accompanying field notes.

Most fundamentally, we rely on the field notes to guide us back to the same locations sampled by Grinnell and his colleagues during their original surveys. The basic goal of the project is to return to the same sites originally surveyed by Grinnell and his teams; this gives the project much of its public appeal as well as maximizing its analytical power. If we conduct our surveys at the same locations, at the same time of year, and with approximately the same field methods as the historic surveys, then our results should be directly comparable. Therefore, any discrepancies between the two datasets are most likely indicative of true changes in species' distributions.

Not surprisingly, some of the most valuable historic records are detailed maps documenting the location and extent of inventory activities at a site. Most of these were contemporary U.S. Geological Survey topographic maps on which Grinnell's field teams marked their daily travel routes or traplines, and which were then bound in the field notes or stored flat in the MVZ's map archives. But even hand-drawn maps in the field notes often provide sufficient detail to effectively locate and replicate the original surveys. One of the most striking examples is the field map drawn by Charles Camp in July 1915 depicting the study site in Lyell Canyon, part of the original Yosemite survey.

Unfortunately, route maps like these were not available for all of the traplines or all of the surveys. For example, we found none from the historic Lassen survey; it is unclear whether these maps were thrown away or were simply never created in the first place. In the absence of these maps, we had to rely on the written location descriptions of the study sites in the field notes.

It may seem surprising that the published monographs did not contain detailed accounts of the individual study sites. Although the species accounts in these monographs generally contain extensive detail about the species' physical appearance, behaviors, and habitat associations, the associated collecting sites are scarcely mentioned. For example, the 594-page monograph for the Lassen Transect survey contains only a single

(*opposite page*) **Hand-drawn map of study site and trap placement in Lyell Canyon, Yosemite National Park, by Charles Camp, July 22, 1915, based on a contemporary USGS topographic map, but with additional details added. Note the topographic contours and streams, and the marks depicting the two traps that captured wolverines (*Gulo gulo*).** Archives of the Museum of Vertebrate Zoology, University of California, Berkeley, http://bscit.berkeley.edu/mvz/volumes.html; Charles L. Camp, 1914–1922, Section 3, page 484.

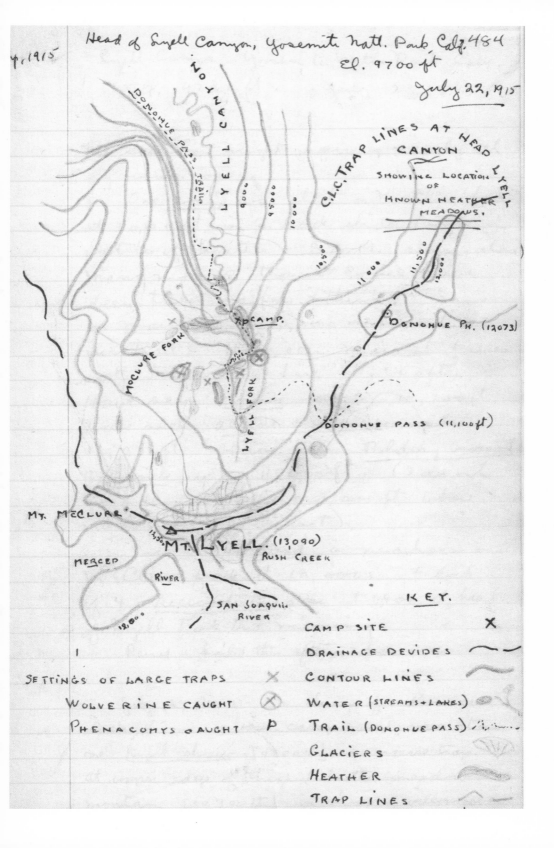

Head of Lyell Canyon, Yosemite Natl. Park, Calif. 484
El. 9700 ft

July 22, 1915

C.L.C. TRAP LINES AT HEAD
CANYON
SHOWING LOCATION
OF
KNOWN HEATHER
MEADOWS.

LYELL CANYON

DONOHUE PASS TRAIL

9000
9500
10000
10500
11000
11500
11500
12000

XP CAMP.

McCLURE FORK

LYELL FORK

DONOHUE PK. (12073)

DONOHUE PASS (11,100 ft)

MT. McCLURE

13300 MT. LYELL. (13,090)
RUSH CREEK

MERCED
RIVER

SAN JOAQUIN
RIVER

12,000

KEY.

CAMP SITE   X

SETTINGS OF LARGE TRAPS   X   DRAINAGE DEVIDES

WOLVERINE CAUGHT   ⊗   CONTOUR LINES

PHENACOMYS CAUGHT   P   WATER (STREAMS + LAKES)

TRAIL (DONOHUE PASS)

GLACIERS

HEATHER

TRAP LINES

y, 1915

map of the entire 3,125-square-mile transect, with the forty-five major collecting sites represented as single dots and summarized only in a cursory table. Moreover, some of these dots were actually aggregations of several individual collecting sites within the immediate vicinity, some of which were sampled months or even years apart. It is possible that Grinnell and his coauthors considered these specific collecting sites to be merely samples of the broader ecological community types being studied, rather than important locales in their own right.

As with most museums, every MVZ specimen bears a tag that includes a succinct description of the location where that specimen was collected, such as "Yosemite Valley, Yosemite National Park, Mariposa County, California." Nevertheless, these brief locality names were often too general for us to relocate the exact site or habitat associations where a particular specimen was collected. Yosemite Valley, for example, extends for more than six linear miles and contains a heterogeneous landscape of sheer granite walls, jumbled boulder fields, dense oak and pine forest, dry and wet meadows, a rich riparian community along the Merced River, a large fen, and at least one lake. Given the habitat specializations of many species, the overly generalized "Yosemite Valley" locality is simply inadequate to document either the location or the ecological context in which a particular species was found.

The problem of overly general locality names was by no means limited to the Yosemite survey, as evidenced by the specimens collected by Joseph Dixon at Eagle Lake in Lassen County in 1921. Eagle Lake is a prominent lake that is easily found on any Lassen County map, but its prominence is a product of its size—more than thirty-five square miles. It is one of the largest lakes in California. Its various bays and drainages contain a wide variety of microhabitats from tule marshes to lava fields. Moreover, the lake falls along a major ecological transition zone: its western shores are dominated by the yellow pine forests typical of the eastern slopes of California's Cascade Range, but along its eastern shores are the sagebrush and juniper of the Great Basin. This ecological transition is reflected in the ranges of several of the vertebrate species in the region. So how to resurvey? Sampling all the possible microhabitats near the lake would be impractical, and selecting new sites randomly or arbitrarily would provide a false comparison with the historic surveys. Fortunately, the field notes that accompanied the specimens collected at Eagle Lake and in Yosemite Valley gave additional details that allowed our team to replicate many of the historic traplines to within a few hundred meters.

The value of the field notes was particularly evident when the locality information on the specimen tag was simply incorrect. For example, MVZ researchers Adrey Borell and Richard Hunt transposed the names of a pair of alpine lakes while collecting in the high elevations of Lassen Volcanic National Park in 1924. Their specimens of American pika (*Ochotona princeps*) and the now-endangered Cascades frog (*Rana cascadae*) attributed to "Lake Helen" were actually collected at nearby Emerald Lake, which has considerably different lakeshore topography and vegetation. Fortunately, Hunt's field notes contained a photograph of "Lake Helen" that made the error obvious.

Later that month, while Borell and Hunt were working out of "Kelly's hunting camp" just outside the park's southeastern boundary, they set one of their traplines at the Drakesbad Resort about two miles inside the park. For reasons known but to themselves, Hunt and Borell assigned the locality of "Kelly's" to these specimens, which were actually collected nearly three miles away, in different microhabitats at a different elevation. This discrepancy would never have been noticed based solely on the information on the specimen tags, but thanks to the detailed daily accounts in their field notes, it was obvious and easily corrected.

The species lists and observations in the field notes also provided records for species that were never preserved as voucher specimens. One of Grinnell's main goals was to document the entire vertebrate community at each study site, but not every species could be collected by trap or gun. This was especially true for many bird species, which could be easily identified by ear or by sight but were not necessarily as easily collected. In the absence of specimens, the field notes provide the only evidence of the presence of these species at these sites. In the Yosemite survey, for example, the field notes contain three times as many bird species than were collected as voucher specimens. In some cases, Grinnell and his colleagues documented these species during timed walks along a particular route or trail in a format similar to a modern timed point count.

These observation records are particularly valuable because they indicate relative abundance and breeding status of some of the

Lake Helen, 8500 ft., S. base Lassen
Peak, Shasta Co.
- Trees are Mountain Hemlock (Tsuga mertensiana)
- Narrow border of vegetation on opposite
  shore is Alpine Heather (Bryanthus breweri)
- Sparse chaparral on upper half of slope is
  stunted manzanita and Holodiscus discolor
- Conies lived in the slide
- Lassen peak in background.

11944

Photograph of Emerald Lake in Lassen Volcanic National Park taken by Richard Hunt in July 1924 and bound in his field notes. Lassen Peak is in the background; marks to the left of the peak are damage on the original print. Hunt mistakenly thought this was the nearby Lake Helen, which lies between Emerald Lake and Lassen Peak. Hunt gave the locality of "Lake Helen" to all the specimens he collected here, including pika from the talus slope at the far side of the lake and several now-endangered Cascades frogs. Archives of the Museum of Vertebrate Zoology, University of California, Berkeley; MVZ image number 11944.

Emerald Lake in 2006.
Photograph by John Perrine.

Storer - 1915   522

Porcupine Flat - June 28

Our joint bird census is as follows:

Species	6:55 8:00 am	8:00 9:00	9:00 10:00	10:00 10:30	11:45 am 12:05 PM	12:05 1:00	1:00 1:45 PM	
Mountain Quail	2	—	1	—	—	—	3	4
Sierra Grouse	2	—	—	—	—	—	—	2
Cabanis Woodpecker	+	+	—	—	—			1
Williamson Sapsucker	1	1	—	—	? 1	1	—	4
Red-shafted Flicker	—	—	—	—	—	—	1	1
Olive-sided Flycatcher	—	1	—	—	1	1	—	3
West. Wood Pewee	1	4	1	—	—	1	1	8
Blue-fronted Jay	1	3	—	—	—	—	1	5
Clark Nutcracker	+	+	—	—	1-2	—	—	3-4
Cassin Purple Finch	—	—	—	—	2	1	—	3
Pine Siskin	—	—	2	1	5	9	1	18
West. Chipping Sparrow	1	2	1	—	2	2	1	9
Sierra Junco	2	2	2	$1+\frac{n}{4}$	1	—	4	12
Lincoln Sparrow	1^{n}_{5}	—	—	—	—	—	—	1^{8}
Thick-billed Fox Sparrow	2	3	—	—	—	—	2	5
Green-tailed Towhee	2	2	—	—	—	—	—	2
West. Tanager	1	3	1	—	—	1	—	6
West. Warbling Vireo	+	4	—	—	—	—	3	7
Audubon Warbler	3	2	3	—	—	1	—	9
Gold Pileolated Warbler	1	1	—	—	—	—	1	3
Sierra Creeper	—	—	1	—	—	1	1	3
Red-breasted Nuthatch	1	2	1	—	—	2	2	8
Mtn. Chickadee	—	4	2	—	—	3	—	9
West. Gold-crown Kinglet	+	2	+	—	—	—	4	7
Ruby-crowned Kinglet	3	4	2	—	1	1	3	14
	17^{11}	42^{18}	18^{12}	2^{2}	14^{8}	24^{12}	28^{14}	

Morning bird census by Tracy Storer at Porcupine Flat, Yosemite National Park, June 28, 1915. Archives of the Museum of Vertebrate Zoology, University of California, Berkeley, http://bscit.berkeley.edu/mvz/volumes.html; field notes of Tracy I. Storer, 1915, Section 1, page 522.

species, and the inventory can be easily duplicated by an experienced birder walking the same path at the same time of day.

The value of such observational records was not limited to birds. The field notes also document the presence of elusive mammals, such as the snow tracks of red foxes and the alarm calls of pikas. Even for small common mammals such as the deer mouse (*Peromyscus maniculatus*), not all of the captured individuals were retained as specimens. Some were irreparably damaged by the snap traps used in the historic surveys, and often more were trapped than was practical to prepare, so many were simply discarded. As a result, the specimens alone are a misleading indication of the relative abundance of that species. Fortunately, the field notes usually contained nightly summaries of the trapping effort and results, indicating the number of traps in the field, the number of individuals of each species captured (often subtotaled by sex and age class), and the number of voucher specimens actually prepared.

The field notes also told us the collecting methodology used in the historic surveys. To be directly comparable, our resurvey should not only locate the same study site but also use comparable methods and effort as the original Grinnellian surveys. The use of different detection methods (mistnets versus shotguns, for example) or profoundly different sampling efforts (multiple trapnights using various types of traps, as opposed to a single afternoon's observations) could confound the comparison. Critical information such as the type and number of traps used and the number of days spent surveying each site simply cannot be recorded on the individual specimen tags due to space constraints, but this information is easily documented in the field notes.

Moreover, the nightly tallies of trap effort and capture success allow a more quantitative comparison than merely whether a species was detected or not. In the past few years, powerful statistical models have been developed that can allow researchers to determine whether a result of "not detected" truly means that the species is not present at the site. These analyses build upon the nightly records of a species' detections, known as that species' "capture history" at the site. Based on these patterns, the occupancy modeling approach can generate per-trap, per-night probabilities of detection for each mammal species.[7] These occupancy models can provide a quantitative statistical answer to whether the failure to detect a species at a site truly indicates that the species does not occur there. Occupancy models have been central to the conclusion that the ranges of several mammal species in the Yosemite region have shifted during the past century, consistent with the expectations of a climate warming scenario.[8] Quite simply, these analyses would not have been possible without the nightly trap effort and capture information recorded in the historic field notes.

Occasionally, the field notes also provide a narrow window of insight into the personalities of their authors. It is important to remember that these field notes were not personal diaries in which Grinnell and his colleagues recorded their private thoughts and aspirations. Rather, the notebooks were institutional records of the distribution, abundance, and ecological contexts of the various vertebrate species observed and collected during each expedition. As a result, the recorded observations were often clinical and detached, which was appropriate given the task at hand. In a few cases, the absence of personal emotion or reflection is jarring. Consider Richard Hunt's notes from July 13, 1924, in Mineral, just south of Lassen Volcanic National Park: "At Mineral we got word that Sam Hermanson, who had been so kind to us at our Lyman Camp [just a few weeks earlier] had been shot dead by two young men who had robbed a bank in Red Bluff and headed for the mountains. The commonest birds by far in Mineral are Robins . . ."

These rare glimpses of the people behind the pens are made more poignant by their brevity, such as Grinnell's mention of building a snowman with his family on Lassen Peak in August 1925, or his occasional checklist of topics to mention in his next letter to his mother. Personal opinions on nonbiological topics were generally not appropriate in the field notes and are rare for most of the field researchers, but Grinnell himself was not above recording occasional expressions of withering disdain. For example, when visited in the field by Joseph Mailliard of the California Academy of Sciences, Grinnell wrote "Mr. Mailliard is well posted in ornithology, but is only a beginner in mammalogy and his curatorship includes mammals as well as birds. He therefore works under some handicap; and his assistants seem to know very little about anything in the natural history line." Grinnell was especially critical of the natural history displays at the newly opened Loomis Museum in Lassen Volcanic National Park in July 1929: "A hodge-podge of poorly mounted birds and mammals, some in travesty of habitat groups, and evidently from all sorts of sources except the Lassen region, is worthless." Grinnell's most vehement criticisms were directed at livestock, which he continually saw as despoiling the fragile high-elevation ecosystems. His anger and frustration are evident in his eloquent recounting of an encounter with cattle inside Lassen Volcanic National Park in July 1925:

> Again I deplore the presence of cattle, which are even, as yet, inside the Lassen National Park. We saw them, stragglers or small herds, at the little meadows in the ravines up the steep slope to fully 8300 feet, a little below (this side of) Lake Helen. They strip the terrane of every edible herb aside from grass, leaving by selection an unnatural flora; they thus open up to dessication [*sic*] by sun and wind every tract of damp drawn-seepage places, where originate the streams lower down; they cut up the ground by trampling, increasing the surface area from which, in the absence of shade from (removed) vegetation, there is increased water loss; they traverse the ridges, leaving trails which, in times of rapidly melting snow or heavy rain, wash and become gullies cutting in to the steep slopes and hastening torrential runoff; they foul the water of the seepage places which are nearly always the sources of the lower streamlets, and they befoul and trample and open up to evaporation the streams themselves . . . As I see the problem, *any* cattle whatsoever are destructive to this high mountain country, scenically, recreationally, florally and faunally, and, most emphatically, from the standpoint of water conservation [emphasis in original].

Reading the handwritten notes of researchers who are long gone can be a profound experience. It is particularly rewarding to see them beginning to develop the ideas that later reached maturity in their published manuscripts. Grinnell mused on the flocking behavior of red-winged blackbirds (*Agelaius phoeniceus*) as he observed them flying along the Merced River at La Grange in 1915. Simple observations could develop into deeper, more general insights. Walter P. Taylor wrote a four-page synopsis entitled "The Ecological Niche" in his notes while working at El Portal in Mariposa County in December 1914, in which he details the basic elements of Grinnell's empirical niche theory as well as those of competitive exclusion. His concluding sentence, written just before his specimen catalog for that day, could serve as a definition of the latter concept in any modern textbook: "Put in another way, the continued existence of a species in a locality where related species are living depends upon the critical differences, slight or large, in the totality of requirements of each."

Similarly, the notes also give a sense of the realities of fieldwork in the early decades of the twentieth century, especially in regard to traveling long distances by automobile. Roads and other infrastructure were poor—in 1921, the 300-mile trip from Berkeley to Eagle Lake took Joseph Dixon four days, including two stops along the way for auto repairs. Four years earlier, while working in the northern Owens Valley, Dixon had broken his wrist cranking the engine of the "Perodipus," the MVZ Model T truck named after the contemporary name of the genus of five-toed kangaroo rats. He did not record the incident in his field notes, but his colleague H. G. White wrote, "Dixon met with accident from the backfire of the Ford and received a broken wrist which sent him back to Berkeley and left the working of the territory to myself." Grinnell frequently referred to the field vehicle simply as "the MVZ machine."

At their field sites, the teams were tireless hikers, scouring the landscape to collect observations and voucher specimens. On a typical field day (August 26, 1915) during his survey of the Yosemite region, while camped at Merced Lake at 7,800-feet elevation, Grinnell's wanderings covered about twelve miles round-trip, with an elevation gain and loss of some 3,000 feet, throughout which he was continually making observations, collecting specimens, and writing notes:

6:45 A.M.—Just in from rounds of traps . . . 8:15 A.M.—Left camp at 7:15, and am now on the Tuolumne Pass Trail up McClure Fork Canyon at about 8300 ft. Trees in sight from this point: Jeffrey pine, red fir . . . 9:15 A.M.—At junction with Isberg Pass Trail, 9000 ft. Surrounded with l-p [lodgepole] pines and red firs . . . 11:35 A.M.—At little lake at 9800 ft near foot of Vogelsang Pass. Trees in sight l-p and mountain pines . . . 2:15 P.M.—After lunching, I went to the largest of the lakes of the vicinity, at about 10,150 ft; it is on a plateau, with scrubby white-bark pines around the west . . . 4:20 P.M.—Back down at 8300 ft, where I heard the cony this morning. My "squeaking" brot an answer, and after ten minutes or so, he appeared and I shot him . . . Coming down into Merced Canyon, I saw a Sceloporus graciosus [sagebrush lizard] at about 7800 ft. Reacht camp at 6:00 P.M.

Daily hikes of twenty miles or more were standard in the original inventories and were conducted in addition to the daily chores of setting and checking traplines and preparing the resulting specimens.[19] There is scarcely any mention of weariness the next day, although after a particularly challenging foray into a deep canyon in the Lassen region in 1924, Richard Hunt noted, "We found plentiful deer sign on the steep and practically inaccessible west slope of the canyon, including several snug beds. I doubt if cattle could have penetrated here and few human beings would be fools enough to do so."

There is no mistaking the fact that these men (for, with the exception of MVZ patron Annie Alexander, her partner Louise Kellogg, and the women who accompanied them, Grinnell's field teams were exclusively male) were phenomenal outdoorsmen and first-rate naturalists. They commonly spent several months in the field at a time, often several times a year. Borell noted dryly, in 1924, "Yesterday PM we went to Chester and took in the rodeo. There were about 2000 people there. This was quite a change after 3 months in the woods."

As we compare their field reality with our own, we can only wonder: How different will ecological fieldwork be in the future? Moreover, what shifts may occur in the cultural values that encompass how our research is conducted and recognized by the public? Will future generations scan our notes for similar historical and cultural clues, in addition to our scientific data? Obviously we have a responsibility to provide the scientific details these future researchers will need to do their work, but the occasional personal anecdote has its merits as well. For example, in the summer of 2006, several students on our Lassen resurvey field team encountered some campers who were angered by our use of mistnets to collect bird voucher specimens in a national forest. When our field team returned to their car at the end of the day, a note was on their windshield:

Brother (and Sister!) Humans—

I can't tell you how much I disapprove of humancentric mist netting! I'm sure you are in love with the science and the ends of greater knowledge about the state of the planet but killing small creatures is unacceptable. These creatures have an inailiable (sp?) right to live free in these forests and you have no right in our National Forest—owned by all to pursue this self serving (doctorate?) end in the name of our species which has gone so cancerous. You compound the problem.

I hope you will let my words be a seed in your consciousness and not defend and placate yourselves with your current rationales.

Peace in the world,

Chris

When the students brought the note back to camp and recounted their story, the current MVZ director Craig Moritz saw the event in a deeper context. Chuckling, he said, "Be sure to bind that in your journal!" We can probably assume that neither Grinnell nor any of his contemporaries received a similar note on their "MVZ machine."

HOW DO WE TAKE FIELD NOTES?

Read above suggestions every few days, devoting half an hour or so to thoughtful consideration of the objects of our field work, which are: To ascertain everything possible in regard to natural history of the vertebrate life of the regions traversed, and to make careful record of the facts gathered in the form of specimens and notes, to be preserved for all time.[20]

—Joseph Grinnell, 1938

An important element of the modern resurvey is establishing a baseline of current conditions to facilitate comparisons in the future. Undoubtedly our field notes will play as important a role in that resurvey as the original notes play in ours. Even outside of the Grinnell Resurvey Project, the tradition of field notes remains alive and well in the MVZ. We still use the same custom-made paper, and some of the corresponding three-ring binders are older than the researchers who carry them. But by no means does every field researcher implement the system in identical fashion. Time, experience, and technology have shaped the way each of us records our information.

Jim's System

I was taught the Grinnellian field note system in graduate school, but writing notes did not become ingrained as a part of my field program until I began a position as assistant curator at the MVZ in early 1969. By the time my own professional career began, the nature of our science had changed. The large, group-oriented field parties targeting general descriptions and collections of local faunas were largely gone, replaced by taxon-specific fieldwork of the individual investigator. As my early research focused on pocket gophers and not on general collecting, my field journal detailed the localities, trapping program, observational records, and other aspects of this specific research program. My journal notes, thus, were both the itinerary component and the species account that came to characterize the MVZ-style notes by the 1930s. I did not write a separate species account then, nor do I now. I did, however, and still do, insert maps of localities and traplines as well as abundant photographs of habitats and animals within my journal, originally gluing or taping both to our standard notebook paper. And, of course, I maintained a sequential numerical catalog of specimens collected and prepared, accompanied by the collecting date, locality, and the standard data taken at the time of preparation—sex, reproductive information based on autopsy, linear body measurements, and so on.

From the earliest days of my fieldwork until now, throughout a given day I jotted notes, typically in pencil, into a small, spiral-bound pocket notebook, remembering the admonition not to trust one's memory but to record observations as continually as possible. I then transcribed these notes into my handwritten journal, using both the archival paper stock supplied by the museum and permanent black ink, in the evenings on the best of days or every few days when an intense field effort allowed. Neither my handwriting nor field circumstances were always optimal, and hurried entries often became difficult for others to decipher. Following the lead of MVZ's third director, Oliver Pearson, who began to type his notes on a computer and print them out on MVZ archival paper in 1990, I began to do the same a decade later. From 2000 onward, I would still take pencil notes in a small pocket notebook in the field, but I transcribed these into a word-processor document with margins set for the size of our field note pages, finally printing this document on unlined MVZ archival paper. I combined this document with my field catalog for a particular trip and eventually both would be bound in the same manner as standard, handwritten MVZ field notes.

This approach had the advantage of producing both an archival paper copy as well as an electronic copy. It was also easy to intersperse specialized maps and digital photographs, which had become the norm by this time, throughout the journal text. The disadvantages of this approach were, at least, twofold. First, notes were not always transcribed until some time after a particular trip, as I don't always take a laptop computer into the field. Thus the chance of forgetting, or even worse, misremembering events or observations was increased. To compensate for this, I now take much more detailed and extensive penciled notes in my small pocket notebook while in the field, referencing these to photo numbers from my digital camera and other relevant information. The second potential disadvantage, and one that only time will determine, is the permanence of the inks used in all commercial printers today, whether laser or inkjet.

Before I began the resurveys of the original Yosemite transect, and during the first several years of that work, I wrote my digital field notes in the same manner as my handwritten notes—that is, daily entries of events and observations in a free form, including details of individual traplines under the date that the line was originally established and details of captures, releases, and specimens retained on each subsequent daily or twice-daily trap check. The information was all there, but one had to read through multiple pages to locate trap records for use in any analytical program, such as the detectability and occupancy modeling we began to employ by 2006. Hence, to facilitate access to these essential data using a slightly modified form developed by John Perrine for his resurvey of the Lassen transect, I began to organize trap data into tabular form with all other relevant information on that trapline physically juxtaposed in the journal with that trap table. Because the resurvey work has focused on particular "sites" where one or more traplines and sets of other observations are gathered over a three- to five-day period, my notes are now organized by those inclusive dates rather than daily, with each trapline detailed as to its georeferenced position, trap type, trap spacing, habitat (including substrate) described in detail, tabular daily trap results, and which captures resulted in voucher specimens (cross-referenced by my field catalog number). At the end of those date sets for that particular "site" are the collective observations I made over those same dates about other organisms encountered as direct observations or indirectly through sign.

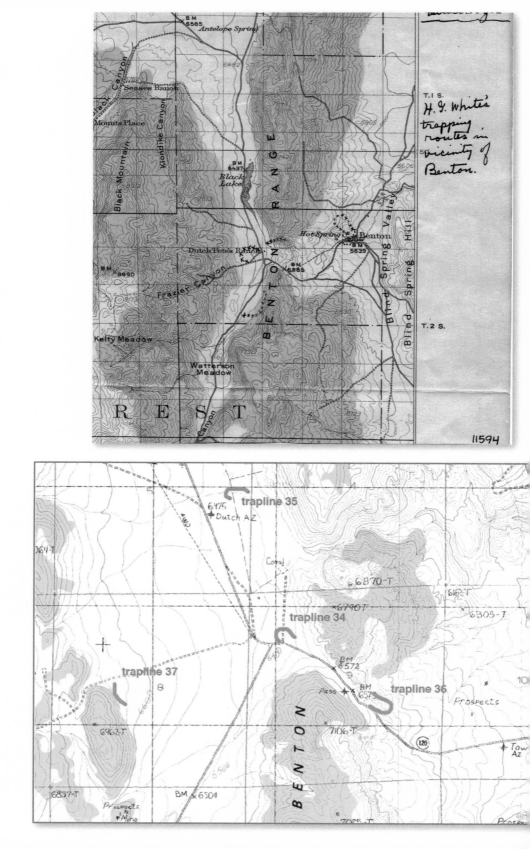

Pages from Jim Patton's field notes, Owens Valley and vicinity trip, eastern California, April and May 2008. (*opposite page, top*) Reproduction of route and trapline map from H. G. White's field notes of September 1917. (*opposite page, bottom*) Current topographic map indicating generalized placement of the four traplines Patton ran during the four-day period of May 16–19, 2008, with locality "site" specified. (*this page, top and middle*) Digital photographs of the habitat of trapline 36. Its position is indicated in the bottom map, opposite. (*this page, bottom*) Details of trapline 36, including dates run, number and type of trap, georeferenced positions of the beginning, middle, and end of the trapline, habitat description, and a table of trap results over the two-day period that line was run.

Patton, J. L.   97
2008
Owens Valley and vicinity – April and May

16 May – 19 May: Trapline 36

TRAPLINE 36: Adobe Valley, Mono Co., California
start date: 17 May 2008   end date: 19 May 2008
trap type and number: 40 Sherman live traps

coordinates: start   –   37.78743°N 118.55948°W   6570 ft
                middle –   37.78746°N 118.55838°W   6590 ft
                end   37.78805°N 118.55970°W   6525 ft

Habitat: Great Basin desertscrub, piñon pine, mountain mahogany, *Ribes* sp. (yellow flower, with spines), wax currant, sagebrush, rabbit brush, bitterbrush, both *Ephedra nevadensis* and *E. viridis*, several species of *Eriogonum*, and Cream bush (*Holodiscus* sp.). The soil is loose and coarse sand within large orange granite outcrops.

This trapline is within the pass crossing the Benton Range between Adobe Valley and Benton Hot Springs in the Blind Spring Valley, on Hwy 120. This is the same habitat but about 0.5 mi S of H. G. White's trapline through the "notch" east of the corral at Dutch Pete's Ranch (see copy of his map on pg. 86).

Table: trap results for Trapline 36

trap #	species	coll? Y/N	field #	traps checked
162	*Peromyscus maniculatus*	yes	24233	18 May am
175	*Peromyscus maniculatus*	yes	24234	"
176	*Neotoma lepida*	yes	24231	
183	*Neotoma lepida*	yes	24232	
190	*Peromyscus maniculatus*	yes	24235	"
195	*Peromyscus truei* juv	no	---	"
162	*Tamias minimus*	yes	24239	19 May pm - pulled
180	*Peromyscus truei*	yes	24240	
183	*Peromyscus maniculatus*	no	---	
189	*Peromyscus maniculatus*	no	---	"

In today's world, where field investigators are using a wide range of electronic devices to record data in as efficient and retrievable a manner as possible so that data can be downloaded directly in a format that requires minimal manipulation for analysis, the system I employ still provides the Grinnellian-era paper record to be preserved indefinitely in the MVZ archives as well as one in which the essential data can be obtained quickly. I have come to like this method both for the way my notes are recorded as well as the method of recording them, and I now use the same approach for all of my field efforts, including those not part of the resurvey project.

John's System

I joined the Grinnell Resurvey Project as a post-doctoral researcher. Although I had been tangentially associated with the MVZ during my dissertation (Jim Patton was one of my committee members), this project was my first immersion into MVZ culture and the Grinnell field note system. In retrospect, I wish that I had used the formal field note system during my dissertation research on the endangered Sierra Nevada red fox—many of the anecdotal observations never made it into my dissertation, much less into peer-reviewed publications, and had I used the Grinnell system these observations would at least have been archived and accessible for the future.

Like many folks at the MVZ, I actually have two field notebooks: a "raw" notebook and my formal Grinnellian notebook. In the field, I take all my raw notes in a waterproof notebook using a fine-point permanent pen (or pencil when it's raining).

The entries have virtually no structure other than the date at the top of (almost) every page; data and observations are jotted down as they come, with whatever abbreviations and notation make sense at the time. This is my private notebook, and so it is the functional equivalent of my short-term memory. At the end of the day, I transcribe the notes into my Grinnellian journal as if I were writing a letter to a colleague. For my catalog, I write directly into the Grinnellian notebook as I prepare my specimens. The curatorial staff at the MVZ frequently refer to our catalogs as they accession and archive the specimens, so I take only the current working page in the field with me and leave the rest back at the

(20 June)   Red Bluff   41

Round of photos @ Mobile Estates woods just, showing
anthropogenic habitat on E side + woodland in W.
Otter swimming from bank to tule island ~ 0845 ~ owner
has seen prev. Kingfisher (pair), wrentit, etc.
Squirrels seen esp. in watered lawn

Junes Ranch: ISm: 40.09311 × 122.22772
                     88m elev   . Start .

T133: Opossum (jv) — water edg.
S132: Microtus —   water's edg.
S139: Mus —   grass .   ┌──────────────────┐
S151: REME ⎫ in concrete   │ S16:   ±7m   │
S152: REME ⎬ amid grass   │ 40.09325 ×   │
                                            │ 122.22687   │
S156: Rattus jv + grassy hillow   │   90m   │
S157: Rattus jr   "   │   elev.   │
S158: Mus   "   └──────────────────┘
S160: Mus   "   8 cardinal photos

                                            end: ±9m 40.09342
                                            88m 122.22513
                                            elev.

    (Σ): 9 caps ; 8 spec. — 1 Microtus, 3 Mus, 2 Rattus, 2 Remy
         1 rel.

Snaps: 5: REME.   DOM: 11AM
_____

7 pm: Bald Eagle soaring over Tule Creek Ranch!
9 pm: Bufo boreas in camp. not collected.

John Perrine's raw field notes in a waterproof notebook, June 20, 2007. They contain trap results, observations, and GPS coordinates. Note the lack of narrative format and the use of abbreviations.

museum. I have never written a species account, but I can see their merits, particularly as part of a focused study on a particular organism or when one is first working in a new region with novel species.

For the first few field seasons I wrote with a disposable technical pen with archival ink, but these pens lasted only a few days and produced a rather grainy line. I recently invested in a refillable technical pen; its line is a bit too frail for field notes, but it is perfect for specimen tags. Unfortunately, technical pens tend to clog if not used regularly, and the cheaper models like mine cannot easily be cleaned. Now that I am a faculty member, I sense a nicer pen in my near future.

I tend not to copy all of the original field data into my journal for fear of transcription errors, and I have not yet acquired Jim's discipline of formatting and printing out my spreadsheets for archival binding in my journal. So it is a trade-off as to how much detail to include in the journal. Rather than recording every capture from every trap (as in my raw notebook), I make daily summaries of how many individuals of each species we captured on each trapline.

When we return from a collecting trip, I print USGS topographic maps of each field site onto unlined archival paper and then use a permanent pen to mark the exact locations of my individual traplines. Although my journal and catalog both contain the GPS coordinates for every trapline, there is simply no substitute for an annotated map to document each trapline's approximate length and path. If we need to return to a site at a later date, such as to assess seasonal or annual variation in species' abundance, these maps ensure that we can set our traps within a few meters of their previous placement.

THE IMPORTANCE OF FIELD NOTES

Yes, you should take notes along with the rest. Yours would be exactly of as much value as anyone's, also any and all (as many as you have time to record) items are liable to be just what will provide the information wanted. You can't tell often in advance which observations will prove valuable. Do record them all![21]

—*Joseph Grinnell, 1908*

J. Perrine
2007

Journal

Red Bluff. Tehama County, CA

20 June — Cooler this morning - prob. 65°F. Pleasant. To Rio Vista trailer park by 0755. See Gray Squirrel running around irrigated lawn. Meadow portion of line got 1 Microtus californicus (Sherman 118). Also a Sceloporus in a Sherman, on "cobbles" half by island. And a Spotted Towhee dead in a Tomahawk. Coyote scat on cobblestones. Back in meadow, Raccoon tracks in dirt road, along with Deer. (and one heard in bushes). Trapline GPS: 40.20260 × 122.21628 ±6m 283 ft elev. [start: 40.20338 × 122.21491; ±7m; 82m elev; end: 40.20233 122.21799 ±7m, 67m elev.] Took one round of 8 cardinal photos (N. NE...) in meadow, another in cobble stone slough, with a few photos of watercourse on each end. Also took a round of photos at Mobile Estates waterfront, showing anthropogenic habitat on E bank, oak woodland (+ mouth of Blue Tent Creek) on W. An Otter is swimming from dense blackberry on our bank to a small island of reeds just offshore. Owner of Mobile Estates said he's seen the otter here before. Saw pair of Kingfishers (one chasing the other). Green Heron flying, hear Wrentit, etc. Again, Gray Squirrel in lawn. Then to June's Ranch. get 9 captures (8 specimens) from 40Sh/10Th: 3 Mus musculus (2 in dry hillside by pasture, 1 in grass by edge of creek); 2 Rattus rattus (both juv; both on dry grassy hillside); 2 Reithrodontomys megalotis (both in pile of broken concrete amid dry grass); 1 Microtus californicus (right at edge of creek) - all in Shermans; all kept. Plus 1 jv. Opossum in Tomahawk at water's edge; released. Snap traps got 1 R. megalotis - pretty well mangled by the trap. Back to camp for lunch + prep. ~1800h - with Chris back to Inks Creek Ranch to activate the Shermans; we set 15 Macabees. Saw a Bald Eagle soaring at Inks Creek, and a Crotalus in rocks w/13 tail buds. At camp, 2100h, found a medium-sized Bufo boreas - not kept. Pleasantly cool + clear.

(left margin, vertical): [GPS: 40.07511 × 122.22772 [wGS-84], ±5 mi; 80m elev]

(20)

John Perrine's official MVZ field notebook entry for June 20, 2007, containing the same basic information as in his raw field notes (see page 243).

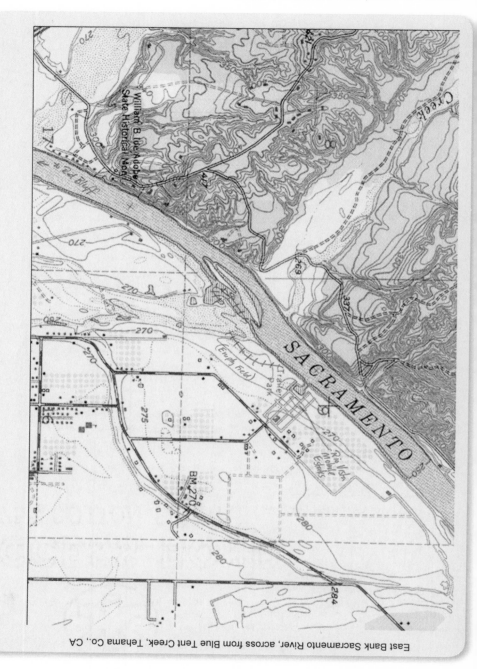

East Bank Sacramento River, across from Blue Tent Creek, Tehama Co., CA

John Perrine's field notebook entry for June 2007, showing an annotated map of a Grinnell re-survey field site referenced in the two previous figures. The header (in the orientation shown, at the bottom) gives the locality for the voucher specimens collected at this site, and the traplines are hand-marked in red ink on a digital version of a USGS topographic map. This map would allow a far more precise replication of the trapline than any written description alone.

Although the historic notes hold a tremendous amount of valuable information, it is no trivial endeavor to compile and extract the necessary details from hundreds of pages of field notes written by more than a dozen researchers. One can easily spend hours poring over page after page, looking for a specific fact or clue. Being an end-user of someone else's field notes certainly gives you insight into the benefits of good note-taking skills. The elation when you eventually find the information you need (or the exasperation when you do not) cannot help but stay in your mind as you write your own notes at the end of a long field day. Our experiences as end-users and creators of archival field notes lead us to a few specific recommendations.

Our first recommendation might be unexpected in a chapter touting the merits of the Grinnellian field note system: Don't get bogged down in the details of format or style. Much has been written about the "rules" for the Grinnellian system, such as the appropriate size of the notebook and margins, the wavy underlining under species names, and the judicious use of space. Such rules are fundamental to any organized collection and, like any style guide, are valuable within an institutional context when excessive variation becomes distracting and inefficient. But these rules are counterproductive if they prevent a researcher from taking field notes in the first place. Field notes are valuable principally because of their content, and although consistent formatting and style make this content more efficient to access, the value of the information is scarcely diminished by taking one's own approach to organization. Similarly, following all the formatting and style rules will be irrelevant if one fails to record the important content. Whether you are starting your first notebook or are an experienced field researcher who wants to improve the quality of your notes, you will get more return by focusing on your content than by refining your formatting.

Second, compose your notes as if you were writing a letter to someone a century in the future. Writing for an external audience requires you to be more explicit in your descriptions and to take less knowledge for granted. Avoid the use of abbreviations, symbols, and other shortcuts that only you will understand. Keep in mind that your primary audience is not you now, but rather someone else later. An archived document is of little use if its content is indecipherable. The goal is to paint a picture of the current context of your work, so that someone else can see the landscape through your words. Now that the era of handwritten communication is nearly dead, we ask our students: How would you describe this to someone over the phone? Ironically, this can be more challenging the more familiar you are with a place or the landscape, because you are more likely to take the details for granted rather than seeing them with a fresh eye. This approach requires a little extra time and attention when writing your notes, but the resulting document will be far more valuable, and not just to others. Few things are more frustrating than not being able to understand your own notes from several seasons ago. Write so that the picture is clear for an external audience, and it will be clearer to you as well.

Writing good field notes takes time and practice, but it need not be onerous. The time required can be greatly reduced if you focus on the key information. Although we cannot know what details may be important in the future—hence Grinnell's direction of "Do write it all down!"—that does not mean that all details are equally important. In particular, avoid the simple diarylike recitation of the day's activities ("Woke up, ate breakfast, checked traplines, prepared specimens, ate lunch, took nap . . ."). In all likelihood, no one really cares whether you ate lunch before preparing your specimens. It is better to spend five minutes writing the important details than twenty minutes writing the trivial ones.

Which details are important depends on the type of work you are doing. If you are collecting voucher specimens, you should record as fully as practicable the ecological context where the specimens were collected and the details of your collecting methodology. This information should be clearly linked to every specimen. Regardless of whether you are collecting specimens, a clear description of your location is almost always valuable: Where are you, and how did you get there? Reliable maps and aerial photos have never been more accessible to field researchers, and there is little excuse for not marking your study sites on one and tucking it inside your notebook. Remember that a picture is worth a thousand words; few written location descriptions are as concise and unambiguous as a mark on a map, even in an era of wristwatch GPS. While there is no reason not to record the GPS points, an associated map can clarify many questions that may arise. A description of the physical structure of your immediate landscape is also likely to be important. Is it heavily forested or dotted with trees? Was there a fire recently, or some other ecological disturbance? Are livestock present? What are the major plant species? Given the current concerns about climate, any indication of the progression of the seasons may be important, such as whether the plants are in flower and what seasonal migrants are present. If you struggle with describing the biota, this may indicate shortcomings in your knowledge of the surrounding plants and wildlife that you may wish to improve upon through field guides, workshops, or other means. Quantitative descriptions are far more likely to be useful than open-ended qualitative descriptions.[22]

Although the ecological questions and research methodologies have changed in the intervening decades, Grinnell's 1938 guidelines for writing field notes remain a valuable reference. Hall included them almost verbatim in his section "Suggestions for Collecting and Preparing Study Specimens of Mammals."[23] Although Grinnell's original memo is not currently available to the public, the MVZ website at the University of California, Berkeley, contains Miller's 1942 revision of these guidelines, along with additional tips on how to take archival field notes. Moreover, in an ongoing effort funded by the National Science Foundation, the original field notes by Grinnell and his contemporaries are being scanned and made available via the Internet as well. These digital copies make the historic field notes accessible to the global community of researchers and also provide an electronic backup should the originals be lost or damaged.[24]

A NOTE ON PERMANENCE

Don't trust your memory, it will trip you up, what is clear now will grow obscure; what is found will be lost. Write down everything in full; time so spent now will be time saved in the end, when you offer your researches to the discriminating public. Don't be satisfied with a dry-as-dust item: clothe a skeleton of fact and breathe life into it with thoughts that glow; let the paper smell of the woods. There's a pulse in each new fact; catch the rhythm before it dies.[25]

—*Eliot Coues, 1874*

Obviously not every field biologist works for an institution that will permanently archive his or her field notes. But that does not mean that a well-documented personal field record is a waste of time. Notes that are thorough yet concise are more likely to be useful for yourself, or your peers or colleagues, and are more likely to be archived as a valuable reference for the future. This is true whether your work consists of biotic surveys, behavioral observations, experiments, or other field pursuits. Our world is very much in flux, perhaps now more than ever before, and long-term datasets are particularly valuable and particularly scarce. Few data collection approaches require as little effort, equipment, or expense to record and archive information as a simple pen and paper. As technologies change with an ever-increasing pace, ink on paper will remain a stable long-term storage medium. It may be challenging to access computer files that are more than a decade old—remember floppy disks?—but a paper record can speak across generations. Write your field notes with this goal in mind.[26]

Why Keep a Field Notebook?

ERICK GREENE

I STUDY ANIMAL BEHAVIOR and ecology, and I have kept field notebooks for as long as I can remember. My field notebooks contain data connected with a specific project I am working on as well as a hodge-podge of miscellaneous observations, questions that come to mind, notes to myself, and descriptions of interesting natural history. These field notebooks are crucial for my research projects. I find that they are also the main source of ideas that takes my research in new directions. I often return to them for the pure enjoyment of reliving special field experiences in some amazing corners of the earth. I can crack the cover of an old field notebook, and these time machines instantly transport me back to watching squadrons of macaws and parrots flying in at dusk to roost in palm swamps in Peru, listening to the "wahoo" alarm calls of olive baboons in the Okavango Delta of Botswana as they warn each other of approaching lions, observing teenage male sperm whales flip their tales up as they begin their hour-long dives to catch giant squid in a deepwater trench off New Zealand, or watching tens of thousands of migrating harp seals, belugas, narwhals, bearded seals, and a mother bowhead whale and her baby stream under arctic cliffs to their summer feeding grounds in Lancaster Sound.

The keeping of field notebooks was considered an absolutely essential activity of the naturalists and scientists of the eighteenth and nineteenth centuries. Indeed, during the giddy heyday of European exploration of the far reaches of the world, the field journals of many scientists and naturalists returning from far-flung expeditions were published and often became bestsellers. Even after a century or more, the field note-books of Maria Sibylla Merian, Thomas Jefferson, Meriwether Lewis and William Clark, John James Audubon, Charles Darwin, Alfred Russel Wallace, Henry Walter Bates, and Henry David Thoreau, to name a few, offer fascinating glimpses into the science and times of these naturalists, explorers, and scientists. Many of these field notebooks are also treasure troves of information we can use to compare the current distribution and abundance of plants and animals.

Since the use of field notebooks in the natural sciences has been so important, I recently started a field notebook assignment for my upper-level Ecology class at the University of Montana. I asked my students to pick one "thing" and observe it carefully over the entire semester. The "thing" they chose could be anything from a single plant, one place, a beaver dam, their garden, a bird feeder, and so on. They had to record their observations at least once a week in a field notebook. One of the main things I wanted to get across is that one of the hardest parts of science is coming up with new questions. Where do fresh new ideas come from? Careful observations of nature are a great place to start. So in addition to their field notebooks, the students also had to suggest at least ten research questions inspired by their observations. This project was worth a considerable portion of their grade for the class, and I thought that this assignment would be received enthusiastically. I could not have been more wrong! After I described the project, the initial chilly response from the students gave way to much rolling of eyeballs and gnashing of teeth. When I asked students about their responses, I got answers such as: "I am interested in science—not creative writing." "This is so lame—I already did my 'expressive arts' requirement." "Do you want us to meditate and write about that too?" Over the course of the semester I noticed a general thawing of attitude about the project, which then grew into a real passion on the part of many students. Here is one representative response written at the end of the project by Carrie Douglas, who observed a single box elder tree in her backyard:

> I have never taken a botany class before and surprisingly I don't know
> a lot about trees from my regular biology classes. This assignment gave
> me an opportunity to ask questions and find answers about trees that
> I have never thought about before. I have never wondered about the
> importance of leaves changing colors in the autumn—what actually

happens during the process, why it is important, and why leaves senesce.

This project also made me learn some important things about myself. I have always thought of myself as entirely left-brained. I love science, procedures, hard facts, etc. I hate abstract, creative or imaginative things. This caused me to be predisposed against this project at the beginning—you mean for my biology class I have to draw pictures and creatively write about a tree all semester!? Give me a break! I thought I was going to dread each journal entry. In reality, I quickly enjoyed making the time—even if it was only 15 minutes or so—to just be outside walking around and quietly observing. It would be the first time in the day that I would not be thinking about school, homework, tests, or the numerous other items on my never-ending "to do" list.

I also really enjoyed just writing informally. Just "throwing up" my thoughts on the page—no worrying about grammar, syntax, proper science writing—none of that—just writing. This is a type of writing I have never been into, but after this assignment I really think I might start keeping a field journal.

After paying attention to all the amazing biological processes unfolding right outside my back door I also realized that I have to live somewhere the seasons change. I took all the amazing seasonal changes for granted. Now I can't imagine living somewhere where the leaves don't change, the snow does not fall, or there are no spring showers. This project has truly caused me to appreciate the beautiful area we live in and I will never look at that box elder tree the same way. Now I wonder what is different for it every time I come home.

I had been puzzled with my students' initial negative reaction to being asked to keep a field notebook for an ecology class. This motivated me to look a bit more deeply to see if this is a general sentiment. To get a sense of the scope, I informally polled many of my colleagues in a broad range of biological fields at several universities. I asked undergraduate and graduate students and faculty how they use notebooks to help record their scientific activities. This much is clear from this poll: in general, lab scientists working in the areas of biochemistry, cell, and molecular biology tend to keep much better notebooks documenting their science than field biologists working in the areas of ecology, behavior, and conservation biology.

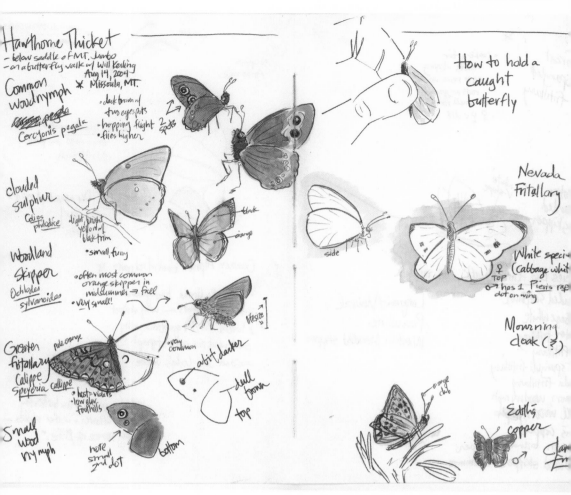

Notebook pages from artist and naturalist Claire Emery describing her observations of butterflies in a hawthorne thicket.

Courtesy of Claire Emery.

Most of my colleagues who conduct laboratory research not only keep extremely thorough and complete lab notebooks, but they teach their students how and why to keep data in them. Some buy hardbound notebooks by the case and distribute them to students working in their labs. They show their students examples of good notebooks and lay out their expectations about how data need to be recorded. They often have review sessions where they look over the students' lab notebooks. Some labs even have friendly competitions in which prizes for the best-kept notebooks are awarded. Many undergraduate courses in microbiology, biochemistry, and molecular biology require students to keep lab notebooks, which can constitute

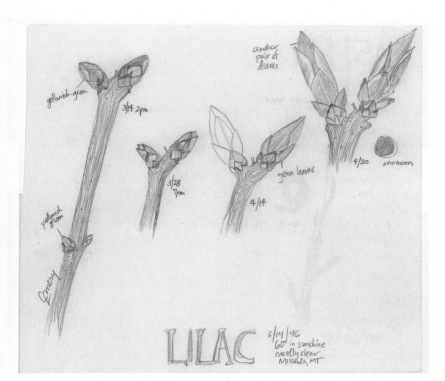

Claire Emery's sketches of the same lilac twig made over thirty-seven days.
Courtesy of Claire Emery.

a substantial portion of the students' grades. Aysha Divan nicely outlines standards for notebooks in laboratory biology.[1]

In stark contrast to the vibrant culture of keeping lab notebooks in molecular biology, my informal polling suggested a lack of interest in notebooks in field biology. When I asked undergraduate and graduate students in ecology and behavior how they kept field notebooks and where they learned to do it, I mainly received blank stares. These were typically followed by responses such as: "I have a GPS for that." "My data are in a spreadsheet." "I write things down when I get home." "I have a computer." The general consensus seemed to be that field notebooks are quaint, archaic, and obsolete in field biology.

In this chapter, I make a plea for the resurrection of notebooks in field biology. I outline the different purposes and functions of notebooks, describe their incredible value to both the authors as well as others, and provide a list of suggested "best practices" for keeping field notebooks.

THE PURPOSE OF
FIELD NOTEBOOKS

People keep field note-
books for a wide variety of
reasons. At one end of the
spectrum, many keep per-
sonal journals that docu-
ment their observations
and experiences in the
natural world. These field
notebooks are close in format and spirit to the notebooks of the naturalists of
the eighteenth and nine-
teenth centuries. This type
of nature-journaling is
mainly pursued by the am-
ateur naturalists—meant
in the best sense—who
love to study and experi-
ence nature without be-
ing paid. These notebooks
tend to capture the beauty
and wonder of the natural
world and help to hone
the observational skills of
the authors. They typically
combine field sketching
and painting with keen
observation. This form of
nature-journaling is thriv-

ing; many museums, natural history societies, and summer camps offer
clinics on how to keep nature journals. For example, the nature artist
and biologist John Muir Laws teaches clinics on field journaling, and his
book and website contain valuable advice and pointers.[2] Other excellent
books, such as those by Hannah Hinchman and Claire Walker Leslie,

October 26, 2004
~11:30 am
clear sky, hazy blue
~45° F, crisp air
N. Hills walk off
Dickinson St.

grunty chortle squeak!
This flock of wide-winged,
short-tailed birds surprised us &
popped up out of the grasses on the
E side of the N. hills.
   Their bodies were mostly dark,
with lighter areas in the primaries on the wing.
   —dark   After seeing them once, we flushed them
   —light   again by walking towards them.
            With the same
chortle-grunt-squeak, they flew up the
hill out of sight.

Note the
rusty-tail

Questions remain:
Later, it seems
we saw a flock   What kind of birds?
of Grey Partridge Habitat needs?
They are common
on the N. Hills   Coloration?
& Mt. Jumbo
Introduced from   Nesting behavior?
Europe — no info
on eggs yet —   Egg coloration?

Meanwhile, magpies chatter in the trees below,
wheels spin & rumble by, flocks of black birds
turn white in the light as they turn. At last,
to sit in silence, to be without motion, and
hear the world offer up its
stories . . .

Perdix perdix
The Grey Partridge

Emery

Claire Emery's observations of bear-marking on a tree (*left*), **and observations and questions about the grey partridge (*Perdix perdix*)** (*above*). Courtesy of Claire Emery.

also focus on integrating observation and art in nature journals.[3]

At the other end of the spectrum, some keep notebooks that follow a much more structured format, such as the Grinnell system. These represent a formal mode of recording when and where specimens were collected. Although they contain scientific data collected in a very formalized way, they lack the personal observations, musings, hypotheses, and sketching typical of nature journals.

To my mind, the most useful and interesting notebooks of field biologists are hybrids; as well as recording details and data of field research, they record the observations, thoughts, musings, and peregrinations of the author. In general, scientists take a range of different approaches in recording their fieldwork because it is a rich and varied undertaking.

WHAT A FIELD NOTEBOOK MEANS TO YOU

A field notebook serves as the basic documentation of your science. The central function of a field notebook is to record and organize your data, and this is the best place to have a complete and accurate record of your experiments and observations. You will find that when you get around to writing up your research, a well-kept field notebook will make that task immeasurably easier. This is because a well-organized field notebook serves as a "central command center" that allows you to collect many other related materials. For example, you may generate lots of information that is not stored in your notebook, such as photographs, recordings, samples collected in the field, and computer files of various types. You can very efficiently organize all of your data in your field notebook. Before embarking on a field project, it is useful to consider what kind of records will be important to you once you complete the work.

Another value of field notebooks is their ability to serve as an incredibly fertile incubator for your ideas and observations. By jotting down interesting observations, questions, and miscellaneous ideas, your field notebook can serve as a powerful catalyst for new experiments and projects. Excellent examples of these sorts of field notebooks can be seen in the work of Bernd Heinrich, Jonathan Kingdon, and many others.[4]

Finally, a well-kept field notebook will give you great pleasure. How quickly we forget! You will find that rereading your field notebooks will give you a chance to revisit corners of nature and remind you of the sorts of natural events that are meaningful to you.

WHAT A FIELD NOTEBOOK MEANS TO OTHERS

Well-kept field notebooks can be extremely valuable sources of information to other people. For example, Henry David Thoreau is best known for *Walden*, which resulted from the field notebook he kept while he lived for two years in his cabin on Walden Pond near Concord, Massachusetts. Until recently, the thoughts and information Thoreau recorded in his notebooks were influential mainly for their social commentary and observations of American society at the beginning of the Industrial Revolution. Thoreau was an excellent naturalist and a keen observer of nature, and between 1851 and 1858 he kept detailed records of the flowering times of about 500 species of plants. His meticulous observations are now extremely valuable, since they were made just before massive amounts of greenhouse gases would be pumped into the atmosphere. Ecologists Charles Willis, Brad Ruhfel, Richard Primack, Abraham Miller-Rushing, and Charles Davis have teamed up to compare Thoreau's observations with the plants that are currently found in the Concord area. In addition to Thoreau's notebooks, they have also been able to locate a number of excellent ones kept by other people in the same area.[5] Between 2004 and 2006 they conducted similar surveys and were able to compare their results with Thoreau's: during the intervening 160 years about 30 percent of the species recorded earlier are gone, and about another 40 percent are so rare they will probably not survive much longer.[6]

Another illustration of the scientific value of well-kept field notebooks comes from the Catalina Mountains near Tucson, Arizona. The Finger Rock Canyon Trail is a strenuous hike, gaining more than 4,000 feet in elevation. The trail spans many different plant zones, and about 40 percent of all species of plants in the Catalina Mountains occur in this one canyon. During the past twenty years, Dave Bertelsen has hiked more than 12,000 miles on this trail and has kept meticulous notes on the flowering dates of almost 600 species of plants.[7] Mike and Theresa Crimmins teamed up with Dave Bertelsen to analyze his incredibly rich data set. These notes allowed them to document profound changes in the plant communities over the twenty years Bertelsen kept his field notebooks: about 15 percent of the species have moved up the mountain and are blooming up to 1,000 feet higher than twenty years ago; some species, such as saguaro cacti and ponderosa pines, seem to be suffering high mortality rates after sustained droughts.[8]

A final example of the persuasive power of field notebooks comes from western Montana. Will Kerling kept detailed field notebooks of his observations of butterflies, plants, birds, and mammals over a quarter-century in Missoula, Montana. He documented the emergence dates, flight times, and locations for ninety-six species of Lepidoptera around the city. The city of Missoula was considering buying private land on Mount Jumbo, a hill overlooking the city, to incorporate into a network of open spaces. There were debates about whether public funds should be used to buy private land for the park system. Will's field notebooks were used as evidence of the special biological diversity and richness of Mount Jumbo,[9] and the city bond measure easily passed. Mount Jumbo is now a cherished crown jewel of the open spaces system for Missoula.

There is clearly tremendous value in keeping good field notebooks, but what do these look like in practice? Are there some common features to consider?

BEST PRACTICES

As I have thought about my own notes and talked with a range of scientists and naturalists, I have assembled a list of topics that I ask my own students to think about before starting fieldwork. These are things that I have found work well for me as well as ideas gleaned from other inveterate keepers of field notebooks. Through the process of assembling this list, two main guiding principles have emerged: First, you will forget things far faster than you expect—most people think they will remember details of their observations and studies for longer and better than they actually do; and second, you will not know at the beginning of a study all the things that might be important or interesting—for this reason, it is a great idea to record more information than you think you might need. These two principles should be kept in mind through all of the following suggestions.

Use a hardbound notebook.   Loose paper is the devil's work! There are a wide variety of good notebooks available, and your choice will depend on your purpose and personal preference. I generally use a notebook that has always been readily available in university bookstores, which are sturdy, inexpensive, and come with lined and numbered pages.

A page from Will Kerling's notebook from June 21, 2001, that documents butterfly observations near the Bitterroot River in Missoula, Montana. Courtesy of Will Kerling.

These are the size of letter paper, which some people find too large. Smaller, hardcover field notebooks with waterproof paper are also popular. Some people use a very small notebook that can fit in a shirt pocket in the field, and then transcribe this information into another notebook at home. I find that this adds another step in the process, and so unless you have a really compelling reason, I suggest using just one field notebook. If you plan to include lots of field sketches, you will want to consider unlined, bound notebooks with paper specially made for art and sketching.

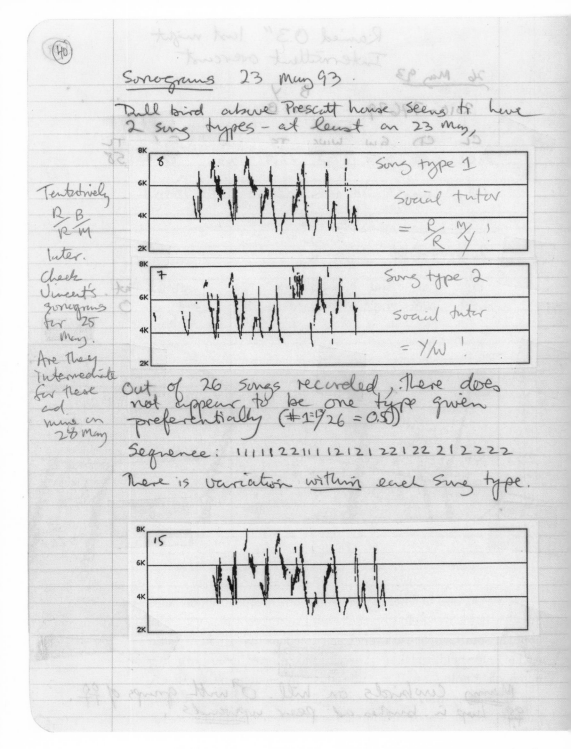

(40)

Sonograms 23 May 93.

Dull bird above Prescott house seems to have
2 song types — at least on 23 May,

Song type 1 — 8
Social tutor
= R/R M/Y !

Tentatively
R/R B/M
tutor.

Check
Vincent's
sonograms
for 25
May.

Are they
Intermediate
for these
and
mine on
28 May

Song type 2 — 7
Social tutor
= Y/W !

Out of 26 songs recorded, there does
not appear to be one type given
preferentially (#1: 14/26 = 0.5)

Sequence: 1 1 1 1 1 2 2 1 1 1 1 2 1 2 1 2 2 1 2 2 2 1 2 2 2 2

There is variation within each song type.

15

My field notes on the songs of a yearling male lazuli bunting (*Passerina amoena*) in Missoula, Montana. Questions raised by these observations led to ongoing studies of song-learning.

Clear + Calm — Am

20 May 93

| | L | R |
| | | |
DD? 2010-77669

CL	CD	GW	WING	TS	TL	WT	FAT
10.38	5.87	4.80	71	17.68	55	320	0

$$320 - 17.0 = 15.0$$

Hump Bird   G/B   M/Y   2010-77668

CL	CD	GW	WING	TS	TL	WT	FAT
10.00	5.70	4.86	71	17.05	56	15	0

Overcast and calm in AM. Started raining at 1300 h, with thunderstorm activity in evening.

I hiked up burn side, but very few of the birds were responsive for netting. We caught a bright bird – DD? and then another bright bird on the hump (= NSF sing 92??). Since got recording s.

Interesting differences in behaviour. Many males are mate-guarding – sticking very close to ♀♀ and not really responding to sings. UB Dull birds seen near fence line gully, singing but skittish – flies away from playbacks. Saw for an UB SY ♂ behind ditch Hutto's house.

Banding records and observations on a young male lazuli bunting with very dull plumage (*Passerina amoena*). These observations and questions led to more studies of plumage that led to another study in 2000.

Keep your contact information in a prominent location. The front of your field notebook is best. In case you lose your notebook, make it obvious how someone can contact you (phone, address, email) and return it to you. I usually offer a small bribe (a small amount of money, ice cream, beer of choice) and lots of good karma.

Write for yourself and for posterity. If nothing else, this encourages me to pay a bit more attention to my lousy penmanship, and increases the chances that even I can read it. Indeed, one of the biggest challenges in using Thoreau's field notebooks has been to decipher his difficult script. This mindset will also encourage you to write very clear descriptions, and minimize obscure references to things that only you will understand. If you keep a good enough field notebook, it will be valuable to others now and in the future.

Write pertinent field information with every new entry. You should enter the date, time, and location at the top of every page. It is a good idea to underline these to highlight them. The "Grinnell system" and other systems have very strict and formulaic ways of recording field information. You may or may not choose to follow these, but at a minimum you should also record elevation, information on habitat type, routes you are traveling, and weather. If your study depends critically on some environmental variable, you will obviously need to record more detailed information on that. For example, if you are studying the songs of tree crickets (whose call rate is extremely sensitive to temperature), you will want to have a very accurate field thermometer with you.

Add information on your location. You should record enough information about where you are working so that another person could return to your exact site. You can do this by including detailed maps, GPS coordinates, sketch maps, and so on. If you are visiting different sites, carefully record information on your routes. If you are conducting intensive studies in just one or a few locations, you can describe those once and then refer back to them. For example, Byron Weber, an avid naturalist in western Montana, kept detailed field notebooks for almost twenty years of his daily observations from one small area along the Bitterroot River. He included a sketch map in his field notebooks and then after that he could refer to specific observations. On December 12, 1983: "4 AM -40°C . . . R'-5 5 magpies, 2 ravens . . . only a sliver of open water near the shore."

A map from Byron Weber's observations of the fauna and flora of the Bitterroot River in Missoula, Montana. Courtesy of Byron Weber.

Record your methods.   When writing about your science in publications, it is important that you include enough methodological details so that someone else could replicate your study. The logical place to record many of these details for field studies is your field notebook. You will forget many of these details faster than you think, and you will search for details when writing your "methods" sections of your papers. It helps to think of using the pages in your field notes as a rough draft of your methods section. If you are collecting information with specific types of equipment or machines, make sure that you record what you are using in your field notebook. For example, if you are recording sounds of insects, the characteristics of those recordings will vary depending upon what type of recorder you are using (tape recorder, digital recorder), recorder settings (for a digital recorder, sampling rate, bit depth, and so on), what type of microphone you are using (shotgun, omni, cardiod, parabola, size of parabola, and so on), filter setting of the microphone, temperature, distance between you and the insect, and what is between you and the insect (lots of vegetation, no vegetation). The interpretation of your data will depend on such details, and so it is critical that you record them accurately while in the field.

Make backup copies.   Hearing even one horror story about someone who did not made copies of field notebooks and lost irreplaceable data is enough to make any field scientist shudder. During the middle of a busy field season, it is well worth spending fifteen minutes at least once a week making photocopies of your new entries. If you are generating lots of irreplaceable data, you may want to make copies more frequently. This is the equivalent of backing up your computer files. You should get in the habit of scheduling this—write it on your calendar and do it. Store your backup copies in a different place from where you keep your field notebooks.

If you use abbreviations, make sure there is a key in your field notebook. Some people are in the habit of using abbreviations to record locations, species, people, and so on in their notebooks. Even you may forget what your abbreviations mean, and certainly it will be confusing or impossible to others trying to interpret your field notebooks. Byron Weber includes a key of the abbreviations he used in his field notebook right next to his field map.

Don't leave home without it. You should make sure you are comfortable with the size and style of your notebook. It is also important that you have something comfortable to carry it in so that there is no temptation to leave it behind. You should feel naked in the field without your field notebook.

Form a writing habit. Writing in your notebook should become second nature to you. Thomas Jefferson was such an inveterate chronicler of daily events in his notebooks that he even took the time to record the weather four times on the day he helped write the Declaration of Independence. So unless you have something far more pressing than writing the Declaration of Independence, you have no excuse for avoiding your field notebook!

Set up a structure for your field notebook. Although many people just start at the beginning of a field notebook and keep a running log, it can be extremely useful to set up sections to keep track of very specific information. You can make tabs that stick out that help you easily find your different sections. For example, at the back of my field notebooks I have found it very useful to include dedicated sections for the following:

Driving log: I keep track of each research-related trip, date, gas and mileage, times left and returned, and destinations. This makes it very easy at the end of the season to summarize all my travel, especially if I have a grant for travel expenses.

Expense log: I record all expenses related to my research. I tape a small envelope right on the page in which to keep the receipts.

Permits: If you need permits or special permission for any of your research, such as banding birds, getting onto wildlife refuges or private property, or collecting rare plants, you should have copies of these with you in the field at all times. It is easy to keep an envelope taped in the back of your notebook with all of your permits.

Photo log: I keep track of all of the photos I take in this log. I record locations and dates, and any notes related to the pictures I will find useful later. You can keep logs for any sort of ancillary information you are generating, such as sound recording logs, sample collection logs, or data logs with the names of computer files you generate.

Contact log: In some of my research I need to get onto private ranches or refuges. I have a list of contact information for ranchers, private land-owners, and refuge managers with whom I need to keep in touch during fieldwork.

You can easily set up your own discrete personalized sections for any sorts of information that will be useful to have collected in one place. Be liberal in the amount of space you leave for each of your sections.

Create an index. An index to your field notebook serves a similar function to the logs described above—they are both very efficient ways for you to organize and find information. They differ in that you can set up a log at the very start, whereas you create an index after (or during) a field season. In an index, you can indicate where in your field notebook you can find information on specific experiments (you may be running several experiments in parallel, and information on them may be inter-leaved in your notebook), specific species of interest, and specific habi-tats or locations. A good index takes some time to compile, but it quickly repays itself by making it very easy for you to locate specific information later. Byron Weber's 1983 notebook shows one page of the index that lists the pages where he made observations for each species. The asterisks indicate pages that contain information on that species more than once on the same page. Byron's indexes contain similarly detailed lists for mammals, reptiles, amphibians, fish, arthropods, plants, the condition of the Bitterroot River (ice, water levels, river traffic), weather, people, astronomical observations, and a miscellaneous section (formation of Bitterroot Audubon chapter, Lee Metcalf National Wildlife Refuge, state legislative actions on wilderness bills and nongame wildlife bill, Bob Marshall Wilderness area, and earthquakes). These are especially de-tailed indexes, but you can see how they make it easy to retrieve infor-mation.

Treat your field notebook like a scrapbook. You should view your field notebook as a central clearinghouse for miscellaneous information that is relevant to your research project. If there are related bits of information that you will find useful later on, sketch them, write them down, photo-copy them and staple or tape them in your notebook. My field notebooks

Birds /182

74. Northern Rough-winged Swallow 67, 74, 75, 85

75. Barn Swallow: 72, 78, 83, 85, 88, 89, 92, 94, 97, 104

76. Cliff Swallow: 88

77. Tree Swallow: 45, 49, 50, 54, 64, 65, 70, 75, 85*

78. Black-billed Magpie: 18*, 21, 22, 23*, 24, 30, 32, 35, 36, 37, 38*, 41, 43, 45, 48, 49, 65, 70*, 73, 74, 75, 76, 77, 79, 80, 84, 85*, 86, 87, 89, 90*, 92, 94, 96*, 99, 100, 102, 105*, 110*, 111, 112, 114*, 115, 120, 121*, 123, 124, 126*, 127, 131*, 133*, 135, 139, 145, 150, 151, 153, 156, 157, 158, 159, 163*, 167, 169, 170, 172, 174*, 176, 177

79. Common Raven: 18, 19, 22*, 23*, 24*, 29, 31*, 32, 34, 38, 39*, 57, 63, 75, 77, 79, 95, 105, 111, 121, 123, 131, 134, 150, 154, 169, 170, 171

80. American Crow: 122, 125

81. Clark's Nutcracker: 38, 60, 120

82. Black-capped Chickadee: 17, 18*, 19, 21*, 23, 27, 29*, 30, 31*, 34, 35*, 36, 37, 38, 41*, 50, 60, 75, 84, 86, 90, 92, 94*, 96, 97, 99, 101, 103, 104, 107, 110*, 111, 112*, 113, 114, 116, 124*, 125, 126, 128, 129*, 131, 132, 133, 134, 137, 140, 141*, 148, 150, 151, 156*, 157, 159, 160, 161, 162, 163*, 167, 169, 172, 174, 176, 177

83. Mountain Chickadee: 51, 69, 93

84. White-breasted Nuthatch: 19, 21, 29, 36*, 39, 51, 74, 75, 92, 94, 102, 115, 126, 128, 134, 140*, 141*, 142, 148, 150, 151*, 170, 174, 177

85. Red-breasted Nuthatch: 19*, 30, 39, 41, 93, 96, 102, 104, 107, 121, 123, 129, 152

86. Brown Creeper 129

87. American Dipper 175

88. House Wren: 65, 66, 67, 70, 73, 75, 76, 84, 85, 86, 87, 89*, 94, 98

89. Winter Wren: 106, 111, 120

90. Marsh Wren: 95

91. Gray Catbird: 75, 94, 95

92. American Robin: 29, 30, 34, 37*, 38*, 40, 46, 50, 59, 61, 65, 69, 70, 74, 75, 83, 85, 86, 88, 89*, 90, 92*, 94, 95, 96, 97, 98*, 99, 100, 102, 104, 106, 107, 111, 112, 114, 148

93. Varied Thrush 98

94. Unidentified Robin 67

Index page from Byron Weber's 1983 field book. Courtesy of Byron Weber.

3 Feb 2007 Saturday
left at 8 AM ed arrived in Kiakura ~ 11³⁰,
Landscape along the way completely hammered - bare
fields, sheep, introduced plts.
Stopped at fur seal haul out.

Sperm whales ~ 18-20 m

Presentation by
Lisa Bond - skipper
with Whale Watch

Kaupapa

Tiaki - guardian
took on killer whale

Tutu

Moko
Toa
patch

Saw
him in
trip

little
nick

Manu = bird

noodle   missing bit

Saddle back - scars on
back

1989 ~ 4000 visitors
last year ~ 1,000,000 visitors
700,000 went whale watching

Notes during my recent research trip to New Zealand documenting the tails of individually recognizable sperm whales in words and sketches.

tend to be repositories for a hodgepodge of what seems like only peripherally relevant material—business cards, newspaper articles, miscellaneous sketches, notes from lectures or articles I have read, ideas to follow up on, and so forth. I am constantly amazed at how often I return to and use the "scrapbook" function of my notebooks.

Two pages from field notebooks I kept while conducting research on lazuli buntings in Montana in May and July 1993. They contain field maps of my study (affixed with tape) along with notes on locations and behaviors of birds seen those mornings, banding information, spectrograms of the songs of individual male buntings, and observations on the behavior of brown-headed cowbirds.

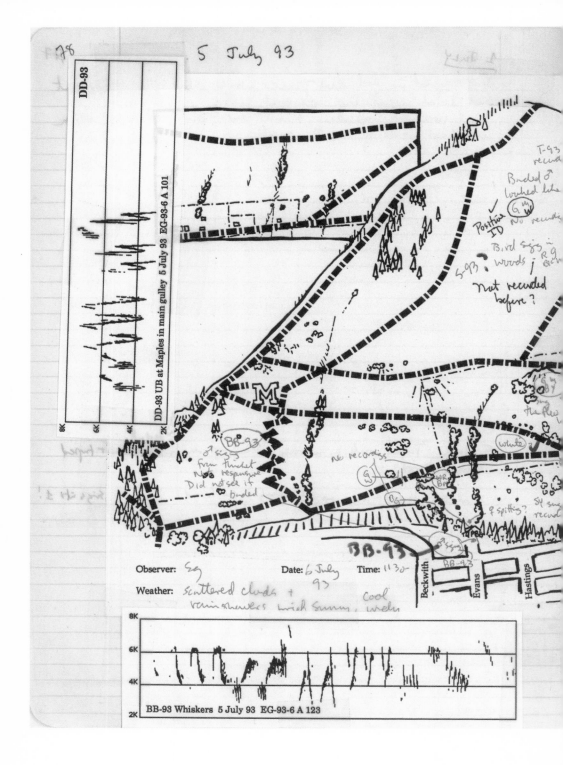

5 July 93

DD-93 UB at Maples in main gulley 5 July 93 EG-93-6 A 101

Observer: Eg   Date: 6 July 93   Time: 1130

Weather: scattered clouds + rain showers with Sunny, with cool

BB-93 Whiskers 5 July 93 EG-93-6 A 123

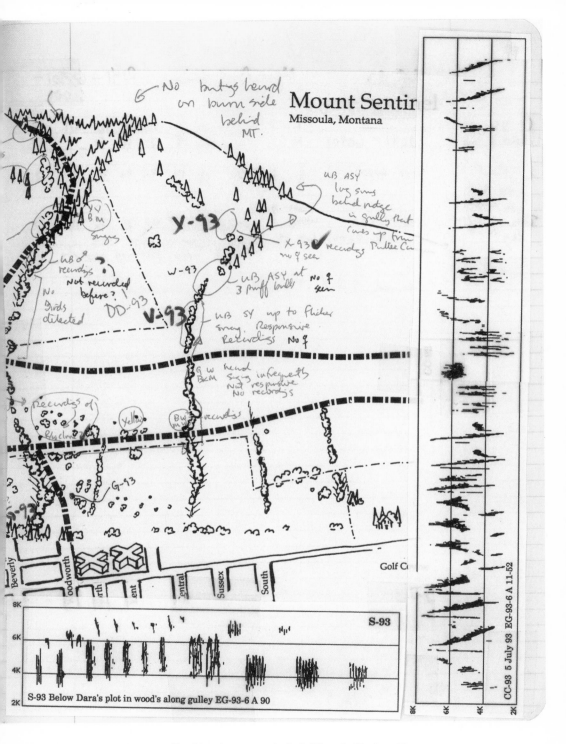

Map showing territory locations of lazuli buntings at a study site in Missoula, Montana. My notes show band combinations, observations on behavior, weather, and songs of certain individual birds.

It is ironic that in spite of the rich history of field notebooks in the natural sciences, this tradition appears to be weakening, especially in the very field that spawned the tradition—field biology. I have made the case that field notebooks are still useful—if not essential—in field biology. This list of suggestions offers a place to start. Decide what your purpose is for keeping a field notebook, and design it in such a way that it works for you. You will find that a well-kept field notebook is a rich form of documentation. It will aid immensely in writing up your research projects, it will be an incredibly fertile place for generating new ideas, it will be a source of pleasure to you as you read it over, and it may prove to be very important to scientists in the future.[10]

NOTES

ACKNOWLEDGMENTS

CONTRIBUTORS

INDEX

NOTES

INTRODUCTION

1. Charles Darwin, *Journal of Researches into the Geology and Natural History of the Various Countries Visited by the H.M.S.* Beagle (London: H. Colburn, 1840). See Richard Dawkins's introduction to *The Origin of Species and The Voyage of the* Beagle (New York: Everyman's Library, 2003) for a discussion of the publication history of the *Voyage*.

2. Darwin, *Journal of Researches*, 468.

3. Richard Keynes, ed., *Charles Darwin's* Beagle *Diary* (Cambridge: Cambridge University Press, 1988); Richard Keynes, ed., *Charles Darwin's Zoology Notes & Specimen Lists from the H.M.S.* Beagle (Cambridge: Cambridge University Press, 2000); G. Chancellor and J. Van Wyhe, *Charles Darwin's Notebooks from the Voyage of the* Beagle (Cambridge: Cambridge University Press, 2009).

4. Keynes, *Charles Darwin's Zoology Notes & Specimen Lists*, 294.

5. Gilbert White, *The Natural History and Antiquities of Selborne, in the County of Southampton: With Engravings, and an Appendix . . .* (London: Printed by T. Bensley, for B. White and son, 1789). The use of "the field" appears in a letter from Lieut.-Col. Montagu to White that was written after Montagu read White's book; Montagu states "I was induced to take this liberty as you say you are a field-naturalist . . ." This letter was published in T. Bell, ed., *The Natural History and Antiquities of Selborne by Rev. Gilbert White*, vol. 2 (London: van Voorst, 1877), 236. Etymological reference from the *Oxford English Dictionary*, 2nd ed., 1989 (online edition). A general discussion of White's Journals can be found in M. E. Bellanca, *Daybooks of Discovery: Nature Diaries in Britain, 1770–1870* (Charlottesville: University of Virginia Press, 2007), 43–77.

6. Linnaeus's Lapland journal was published only after his death: C. v. Linné and J. E. Smith, *Lachesis Lapponica: Or a Tour in Lapland, Now First Published from the Original Manuscript Journal of the Celebrated Linnaeus* (London: White and Cochrane, 1811). A facsimile and thorough discussion of this document is now available:

C. v. Linné, S. Fries, A. Hellbom, and R. Jacobsson, *Iter Lapponicum: Lappl. ndska resan 1732*, vol. 1 *Dagboken*, vol. 2 *Kommetardel*, vol. 3 *Facsimileutgåva* (Umeå, Kungl: Skytteanska Samfundet, 2003–2005).

7. D. Preston and M. Preston, *A Pirate of Exquisite Mind: Explorer, Naturalist, and Buccaneer: The Life of William Dampier* (New York: Walker and Company, 2004).

8. J. Masefield, *Dampier's Voyages: Consisting of a New Voyage Round the World, a Supplement to the Voyage Round the World, Two Voyages to Campeachy, A Discourse of Winds, a Voyage to New Holland, and A Vindication, in answer to the Chimerical Relation of William Funnell*, 2 vols. (London: E. Grant Richards, 1906), 47.

9. See discussion in Preston and Preston, *Pirate of Exquisite Mind*, 3.

10. Banks's notebooks are available online at http://www2.sl.nsw.gov.au/banks/series_03/03_701.cfm (accessed January 2011).

11. R. Spruce, *Notes of a Botanist on the Amazon and Andes* (New York: Johnson Reprint Corp., 1970); H. W. Bates, *The Naturalist on the River Amazons: A Record of Adventures, Habits of Animals, Sketches of the Brazilian and Indian Life, and Aspects of Nature Under the Equator, During Eleven Years of Travel* (London: J. Murray, 1875); A. R. Wallace, *The Malay Archipelago: The Land of the Orang-utan and the Bird of Paradise: A Narrative of Travel with Studies of Man and Nature*, 2 vols. (London: Macmillan, 1869).

12. D. Barrington, *The Naturalist's Journal* (London, 1767). The first edition of this volume was published anonymously. A discussion of Barrington's influence on Gilbert White can be found in P. Foster, *Gilbert White and His Records* (London: Christopher Helm, 1988), 84–114.

13. D. D. Jackson, *Letters of the Lewis and Clark Expedition, with Related Documents, 1783–1854*, 2nd ed., vol. 1 (Urbana: University of Illinois Press, 1978), 62. Journals are available online at http://lewisandclarkjournals.unl.edu/index.html (accessed January 2011).

14. Circular sent to Thoreau from Agassiz entitled "Directions for collecting fishes and other objects of natural history." This circular was inserted by Thoreau in one of his notebooks that he titled "Henry D. Thoreau's Fact-book. Mainly on natural history." The manuscript is held in the Harry Elkins Widener Collection of Harvard's Houghton Library.

15. A. Newton, "On a method of registering natural history observations," *Transactions of the Norfolk and Norwich Naturalists' Society* 1 (1870): 24–32; J. A. Harvie-Brown, "On uniformity of method in recording Natural History Observations, especially as regards Distribution and Migration; with specimen tables of a plan proposed," *Proceedings of the Natural History Society of Glasgow* 3 (1876): 115–123; A. H. Felger, "A card system of note-keeping," *Auk* 24 (1907): 200–205; C. L. Hogue, "A field-note

form for general insect collecting," *Annals of the Entomological Society of America* 59 (1966): 230–233; S. W. Kress, *The Audubon Society Handbook for Birders* (New York: Scribner, 1981), 62–81; S. G. Herman, *The Naturalist's Field Journal: A Manual of Instruction Based on a System Established by Joseph Grinnell* (Vermillion, SD: Buteo Books, 1986); H. R. Bernard, *Research Methods in Anthropology*, 4th ed. (Lanham, MD: AltaMira Press, 2006), 387–412.

16. K. Johnson, *The Sierra Club Guide to Sketching Nature* (San Francisco: Sierra Club Books, 1997); C. W. Leslie and C. E. Roth, *Keeping a Nature Journal* (North Adams, MA: Storey, 2000); N. B. Estrin and C. W. Johnson, *In Season: A Natural History of the New England Year* (Hanover, NH: University Press of New England, 2002); J. New, *Drawing from Life: The Journal as Art* (New York: Princeton Architectural Press, 2005).

17. S. Herbert, ed., "The Red Notebook of Charles Darwin," *Bulletin BMNH (Hist.)* 7 (1980): 1–164.

18. R. B. Yeh and S. Klemmer, *Field Notes on Field Notes: Informing Technology Support for Biologists*, Technical Report, Stanford InfoLab, 2004, at http://ilpubs.stanford .edu:8090/654/ (accessed January 2011); R. B. Yeh, C. Liao, S. Klemmer, F. Guimbretiere, B. Lee, B. Kakaradov, J. Stamberger, and A. Paepcke, "ButterflyNet: A mobile capture and access system for field biology research," *Conference on Human Factors in Computing Systems* (CHI 2006): 1–10.

3. ONE AND A HALF CHEERS FOR LIST-KEEPING

1. K. Kaufman, *Kingbird Highway* (Boston: Houghton Mifflin, 1997).

2. R. C. Stebbins, *A Field Guide to Western Reptiles and Amphibians*, 2nd ed. (Boston: Houghton Mifflin, 1985).

3. L. Jones, "The Lorain County, Ohio, 1898 horizon," *Wilson Bulletin* 11, no. 1 (1899): 2–4.

4. L. Jones, "All day with the birds," *Wilson Bulletin* 11, no. 3 (1899): 41–45.

5. K. S. Brown, Jr., "Maximizing daily butterfly counts," *Journal of the Lepidopterists' Society* 26, no. 3 (1972): 183–196.

6. R. Rolley, Wisconsin Checklist Project 2007, Wisconsin Department of Natural Resources Special Report, 2007.

4. A REFLECTION OF THE TRUTH

1. P. Crowcroft, *Elton's Ecologists* (Chicago: University of Chicago Press, 1991); C. S. Elton, *Animal Ecology* (London: Methuen, 1927); C. S. Elton, *The Ecology of Invasions by Animals and Plants* (London: Methuen, 1958).

2. C. S. Elton, *The Pattern of Animal Communities* (London: Methuen, 1966).

3. C. Darwin, *The Voyage of the* Beagle (London: John Murray, 1839).

4. See www.darwin-online.org.uk (accessed January 2011).

5. R. H. Macarthur and E. O. Wilson, *The Theory of Island Biogeography* (Princeton: Princeton University Press, 1967).

6. R. L. Kitching, *Food Webs and Container Habitats: The Natural History and Ecology of Phytotelmata* (Cambridge: Cambridge University Press, 2000).

7. W. Laurence, *Stinging Trees and Wait-a-whiles: Confessions of a Rainforest Biologist* (Chicago: University of Chicago Press, 2000).

7. IN THE EYE OF THE BEHOLDER

1. J. Kingdon, *East African Mammals: An Atlas of Evolution in Africa*, vol. 1 (London: Academic Press, 1971), v.

2. Kingdon, *East African Mammals*, 2–4.

3. M. R. A. Chance, "An interpretation of some agonistic postures; the role of 'cut-off' acts and postures," *Symposia of the Zoological Society of London* 8 (1962): 71–99.

4. P. Marler, "Communication in monkeys and apes," in *Monkeys and Apes: Field Studies of Ecology and Behavior*, ed. I. DeVore, 544–584 (New York: Holt, Rinehart and Winston, 1965).

8. WHY SKETCH?

1. S. L. Montgomery, *The Chicago Guide to Communicating Science* (Chicago: University of Chicago Press, 2003).

2. Edward Bell, personal communication, January 2008.

3. Lucy Reading-Ikanda, personal communication, January 2008.

9. THE EVOLUTION AND FATE OF BOTANICAL FIELD BOOKS

1. J. E. Graustein, "Nuttall's travels into the old Northwest. An unpublished 1810 diary," *Chronica Botanica* 14, no. 1/2 (1952): 1–88; D. Douglas, *Journal Kept by David Douglas in North America, 1823–1827* (London: W. Wesley and Son, 1914); J. C. Frémont, *Report of the Exploring Expedition to the Rocky Mountains in the Year 1842 and to Oregon and Northern California in the Years 1843–44* (Washington: Blair and Rives, 1845); S. L. Welsh, *John Charles Frémont, Botanical Explorer* (St. Louis: Missouri Botanical Garden Press, 1998).

2. J. K. Townsend, *Narrative of a Journey across the Rocky Mountains, to the Columbia River, and a Visit to the Sandwich Islands, Chili, etc., with a Scientific Appendix* (Philadelphia: H. Perkins, 1839).

3. S. D. McKelvey, *Botanical Explorations of the Trans-Mississippi West, 1790–1850* (Boston: Arnold Arboretum, 1955; reprinted with an introduction by S. D. Beckham, Northwest Reprints, Oregon State University Press, 1997). For information on later naturalists, see J. Ewan and N. D. Ewan, *Biographical Dictionary of Rocky Mountain Naturalists, a Guide to the Writings and Collections of Botanists, Zoologists, Geologists, Artists and Photographers, 1682–1932* (Bohn: Utrecht, 1981); J. L. Reveal, *Gentle Conquest: The Botanical Discovery of North America with Illustrations from the Library of Congress* (Washington, D.C.: Starwood, 1992).

4. J. L. Reveal and J. S. Pringle, "Taxonomic botany and floristics," in *Flora of North America North of Mexico*, vol. 1, 157–192, ed. Flora of North America Editorial Committee (New York: Oxford University Press, 1993). Online at http://www.plantsystematics.org/reveal/pbio/usda/fnach7.html (accessed January 2011).

5. http://www.esg.montana.edu/gl/index.html (accessed January 2011).

6. http://www.plantsystematics.org/tompkins.html (accessed January 2011).

7. See "Index herbarium, Part I. Herbaria of the World," maintained by the New York Botanical Garden, at http://sweetgum.nybg.org/ih/ (accessed January 2011).

11. LETTERS TO THE FUTURE

1. J. Grinnell, "The methods and uses of a research museum," *Popular Science Monthly* 77 (1910): 163–169.

2. Grinnell, "Methods and uses."

3. J. Grinnell, "The niche-relationships of the California thrasher," *Auk* 34 (1917): 427–433; J. Grinnell, "Field tests of theories concerning distributional control," *American Naturalist* 51 (1917): 115–128.

4. F. E. Clements, "Plant succession: Analysis of the development of vegetation," Carnegie Institute of Washington Publication 242 (Washington, D.C., 1916).

5. For Yosemite, see J. Grinnell and T. I. Storer, *Animal Life in the Yosemite* (Berkeley: University of California Press, 1924); for Lassen, see J. Grinnell, J. Dixon, and J. M. Linsdale, *Vertebrate Natural History of a Section of Northern California through the Lassen Peak Region* (Berkeley: University of California Press, 1930); for San Bernardino, see J. Grinnell, "The biota of the San Bernardino Mountains," *University of California Publications in Zoology* 5 (1908): 1–170 + 24 plates; for San Jacinto Mountains, see J. Grinnell and H. S. Swarth, "An account of the birds and mammals

of the San Jacinto area of southern California," *University of California Publications in Zoology* 10 (1913): 197–406; and for the lower Colorado River, see J. Grinnell, "An account of the mammals and birds of the lower Colorado Valley," *University of California Publications in Zoology* 12 (1914): 51–294 + 11 plates.

6. C. Moritz, J. L. Patton, C. J. Conroy, J. L. Parra, G. C. White, and S. R. Beissinger, "Impact of a century of climate change on small-mammal communities in Yosemite National Park," USA, *Science* 322 (2008): 261–264.

7. Craig Moritz, Museum of Vertebrate Zoology (MVZ) Director, personal communication, 2007.

8. J. Grinnell, unpublished letter to Annie M. Alexander dated 18 February 1908, Bancroft Archives, University of California, Berkeley.

9. For example, P. S. Martin and C. R. Szuter, "War zones and game sinks in Lewis and Clark's West," *Conservation Biology* 13 (1999): 36–45.

10. S. G. Herman, *The Naturalist's Field Journal: A Manual of Instruction Based on a System Established by Joseph Grinnell* (Vermillion, SD: Buteo Books, 1986).

11. Herman, *The Naturalist's Field Journal.*

12. E. R. Hall, *The Mammals of North America*, 2nd ed. (New York: John Wiley and Sons, 1981); Herman, *The Naturalist's Field Journal*; J. V. Remsen, Jr., "On taking field notes," *American Birds* 31 (1977): 946–953.

13. E. R. Hall, *The Mammals of Nevada* (Berkeley: University of California Press, 1946).

14. J. Grinnell, "Suggestions as to collecting; note taking; suggestions as to life history notes," unpublished internal memorandum dated 20 April 1938, MVZ Archives (Museum of Vertebrate Zoology, University of California, Berkeley), p. 1.

15. A. H. Miller, "Suggestions as to collecting; note taking; suggestions as to life history notes," unpublished internal memorandum dated 2 July 1942, MVZ Archives, p. 8, revised from Grinnell, "Suggestions as to collecting." Also available online at http://mvz.berkeley.edu/Suggestions_Collecting.html (accessed January 2011).

16. Grinnell, "Suggestions as to collecting."

17. D. I. MacKenzie, J. D. Nichols, J. A. Royle, K. H. Pollock, L. L. Bailey, and J. E. Hines, *Occupancy Estimation and Modeling* (New York: Academic Press, 2006).

18. Moritz et al., "Impact of a century of climate change."

19. K. Brower, "Disturbing Yosemite," *California Magazine* 117 (2006): 14–21, 41–44.

20. Grinnell, "Suggestions as to collecting."

21. J. Grinnell, unpublished letter to Annie M. Alexander, dated 16 April 1908, Bancroft Archives, University of California, Berkeley.

22. Remsen, "On taking field notes."

23. Hall, *Mammals of North America*.

24. They can be accessed at http://bscit.berkeley.edu/mvz/volumes.html.

25. E. Coues, *Field Ornithology* (Salem, MA: Naturalists' Agency, 1874).

26. We are grateful to Barbara Stein, who provided the specifics of Grinnell's 1908 correspondence with Annie Alexander, and to Mary Sunderland, who located Grinnell's 1938 instructions on collecting and field notes in the MVZ archives.

12. WHY KEEP A FIELD NOTEBOOK?

1. A. Divan, *Communication Skills for the Biosciences: A Graduate Guide* (Oxford: Oxford University Press, 2009).

2. John Muir Laws, *The Laws Field Guide to the Sierra Nevada* (Berkeley, CA: Heyday Books, 2007); http://www.johnmuirlaws.com/equipmentlist.htm (accessed January 2011).

3. H. Hinchman, *Life in Hand: Creating the Illuminated Journal* (Salt Lake City: Gibbs-Smith, 1991); H. Hinchman, *A Trail through Leaves: The Journal as a Path to Place* (New York: W.W. Norton, 1997); C. W. Leslie, *Nature Drawing: A Tool for Learning* (Dubuque, IA: Kendall/Hunt, 1995); C. W. Leslie, *Nature Journal: A Guided Journal for Illustrating and Recording Your Observations of the Natural World* (Pownal, VT: Storey Publishing, 1998); C. W. Leslie, *The Art of Field Sketching* (Englewood Cliffs, NJ: Prentice-Hall, 1984).

4. B. Heinrich, *The Trees in My Forest* (New York: Harper Collins, 1998); B. Heinrich, *Winter World* (New York: Ecco, 2003); J. Kingdon, *East African Mammals: An Atlas of Evolution in Africa* (Chicago: University of Chicago Press, 1984); J. Kingdon, *Island Africa: The Evolution of Africa's Rare Animals and Plants* (Princeton: Princeton University Press, 1989).

5. C. Dean, "Thoreau is rediscovered as a climatologist," *New York Times*, 28 October 2008.

6. C. G. Willis, B. Ruhfel, R. B. Primack, A. J. Miller-Rushing, and C. C. Davis, "Phylogenetic patterns of species loss in Thoreau's woods are driven by climate change," *Proceedings of the National Academy of Sciences* 105, no. 44 (2008): 17029–17033.

7. Z. Guido, "Phenology, citizen science, and Dave Bertelsen: 25 years of plant blooms on the Finger Rock Trail in the Santa Catalina Mountains," *Southwest Climate Outlook*, August 2008.

8. T. M. Crimmins, M. A. Crimmins, D. Bertlesen, and J. Balmat, "Relationships between flowering diversity and climatic variables along an elevation gradient," *International Journal of Biometeorology* 52 (2007): 353–366.

9. S. Devlin, *The Missoulian*, 4 April 1993.

10. I thank Paul Alaback, Barry Brown, Claire Emory, Will Kerling, Byron Weber, and Brian Williams for helpful discussions and generous access to their field notebooks. This chapter is in memory of Byron Weber who passed away during this project. He inspired several generations of naturalists with his natural history skills and his field notebooks.

ACKNOWLEDGMENTS

Editing this book brought me into contact with a group of inspiring and accomplished collaborators, and for that I am truly grateful. Part of the challenge of this project was finding eminent scientists and naturalists willing to open their personal field notebooks—documents that were never intended to serve this purpose—to a general audience. The generosity of these individuals has provided an invaluable service to a legion of field scientists and naturalists.

My personal thanks start with my dear wife Jen, who has calmly supported me throughout the duration of this project with good humor, specific and critical advice, and substantial patience. For this steadfast love and support I have dedicated this book to her.

I am fortunate to have had the encouragement of other family members who have helped me bring this project to its completion, including the inspiration and energy of my sons, Riley and Mitchell, and the longstanding support of my parents, Charles and Dorothea Canfield, for my education and research. Thanks to my siblings, David Canfield, Greg Canfield, and Lori Peiffer, for fostering my early love of the field on many family expeditions. Holly Johnston has been a ready and knowing presence and has given freely of her academic perspective.

The colleagues and friends with whom I have worked at Eliot House have provided critical elements to my ability to complete this work. Lino Pertile and Anna Bensted have given more than a decade of their friendship and support of my academic activities. More recently, Doug Melton and Gail O'Keefe have offered similar support. I am also thankful for the assistance and advice of Ree Russell, Lola Irele, Francisco Medeiros, and Sue Welman. I owe a great deal to Emily MacWilliams for her keen eye, unending patience, and her uncommon love of syntax.

The research involved in this book put me in touch with an impressive caste of librarians, and their contributions far exceeded what could have been reasonably expected. I thank the staff of the Ernst Mayr Library, including Mary Sears, Ronnie Broadfoot, Dorothy Barr, and Robert Young. The substantial assistance provided by Fred Burchsted at Harvard's Widener Library is deeply appreciated, and his knowledge of the history of science and field documentation were critical in the development of this project. I also thank Ben Sherwood at the Linnaean Society of London, Auste Mickunaite at the British Library, Earle Spamer at the American Philosophical Society, and Adam Perkins at the Cambridge University Library.

I have truly enjoyed working with the outstanding people at Harvard University Press and have gained much from their experience and counsel. Thanks go to Anne Zarrella, especially for her encouragement and patience, and to Lisa Roberts for taking an ambiguous vision and making it real and beautiful. I am grateful to Kate Brick for showing me how an expert editor actually works, and for the thoughtful attention and energy she dedicated to shaping this book. Thanks also to Rose Ann Miller for helping put this book in the hands of people who might benefit from it. I have sincere appreciation for the time, energy, and honesty of David Foster, Jack Hailman, and an anonymous reviewer whose ideas significantly improved the content of this book. A special thanks to Michael Fisher who believed in the project from the start. He helped navigate challenges and find solutions to problems in a way that revealed a wealth of experience and wisdom.

There are many others who supported me in exploring this project and in developing its underlying idea, from its earliest days, including: Jeanne Altmann, Bruce Archibald, H. Russell Bernard, Andrew Berry, Melisa Beveridge, David Canfield, Lisa Cliggett, Chris Conroy, Sonia DeYoung, Thomas Eisner, David Haig, Gardner Hendrie and Karen Johansen, Brett Huggett, Farish Jenkins Jr., Christin Jones, Robin Kimmerer, Scott Klemmer, Conrad Kottak, Mike Overton, Shawn Patterson, Mark Sabaj Perez, David Pilbeam, Peter Raven, James V. Remsen, Andrew Richford, Ben Roberts, Gary Rosenberg, Michael Ryan, Andy Spencer, Robert Stebbins, David Stejskal, and H. Todd Swimmer. Special thanks to Kathy Horton for many conversations about notebooks, and for her appreciation of good lunch conversation and fine chocolate.

John W. Gruber has listened, suggested, reflected, and given freely of his perspective during the entirety of this project, and I feel lucky to have such a longstanding intellectual companion.

My final and sincere thanks go to Naomi E. Pierce. Without her seemingly boundless investment in me, I literally would not be where I am now, and this book would simply not exist. I have never met a person as incisive and intelligent who is also so warm, understanding, and accepting. Being involved in the community of scientists she has created, and benefiting from her perspective and advice, have been two of the greatest fortunes of my life.

CONTRIBUTORS

ANNA K. BEHRENSMEYER is Curator of Vertebrate Paleontology in the Department of Paleobiology at the National Museum of Natural History, Smithsonian Institution. Her research in Africa, Asia, and North America investigates diverse questions in human evolution, vertebrate taphonomy, paleoecology, and the evolutionary impact of climate change.

MICHAEL R. CANFIELD's work ranges over diverese terrain in science and natural history, and includes empirical investigation of alternative developmental forms in the caterpillars of New World emerald moths. He also studies the history and process of scientific documentation, and teaches courses on camouflage and mimicry, and the biology of insects, in the Department of Organismic and Evolutionary Biology at Harvard University.

ERICK GREENE's studies on diverse organisms from caterpillars to cowbirds empirically test evolutionary hypotheses in the field. He is Professor of Biology and Wildlife Biology at the University of Montana. Some of his field sites are literally steps from his lab, but others take him to remote locations in Costa Rica and New Zealand.

BERND HEINRICH has investigated diverse research topics over the course of his career, including insect thermoregulation, animal behavior, and cognition. He has published over fifteen books on biology and natural history including *Bumblebee Economics*, *Thermal Warriors*, *Why We Run*, and *The Mind of the Raven*, for which he was awarded a John Burroughs Medal. He is Professor Emeritus at the University of Vermont and is also an award-winning marathon and ultramarathon runner.

KENN KAUFMAN started birding as a young child, and he is now one of the most accomplished and revered birders in the world. He is a field editor for *Audubon* and a columnist for *Birder's World* and *Bird Watcher's Digest*, and has written *Lives of North American Birds* as well as five volumes in the Kaufman Field Guide series. He described his early hitchhiking travels and his 1973 "big year" in *Kingbird Highway*, and has recently released a second memoir, *Flights Against the Sunset*. He currently resides in Oak Harbor, Ohio.

JENNY KELLER is a science illustrator whose work has been featured in a number of books and periodicals, including *Scientific American* and *National Geographic*. Her illustrated field journals have been exhibited in several venues, including, most recently, New Mexico and Portugal, where her work from a recent trip to the Amazon was presented. Keller is also an award-winning teacher, and is co-founder of the Science Illustration graduate program at California State University Monterey Bay.

JONATHAN KINGDON is a zoological illustrator, ecologist, writer, and Senior Research Associate in the Department of Zoology at the University of Oxford. A world authority on African mammals and scientific illustration, he has written many books, including *Mammals of Eastern Africa*, *Field Guide to African Mammals*, *Island Africa*, and *Lowly Origins*.

ROGER KITCHING is a tropical ecologist and Professor of Ecology at Griffith University in Brisbane, Australia. His research focuses on insect ecology, biodiversity, ecosystem management, and canopy science, and his fieldwork has been conducted in remote locations in India, Papua New Guinea, and Brunei. He has authored numerous scientific papers and eleven books, including *Food Webs and Container Habitats: The Natural History and Ecology of Phytotelmata*.

KAREN L. KRAMER is Associate Professor in Human Evolutionary Biology at Harvard University and studies traditional forager and agriculturist communities in Mexico, Venezuela, and Madagascar. Her work addresses how these subsistence lifestyles affect the pace of human reproduction as well as how they are inevitably integrated into larger economies. In addition to her many contributions to the primary literature, her recent book *Maya Children: Helpers at the Farm* investigates the interaction between children's labor and fertility in an isolated village on the Yucatán Peninsula.

PIOTR NASKRECKI is a world expert on katydids who is a Research Associate with the Museum of Comparative Zoology at Harvard University. Aside from being the author of many scientific papers, he is the developer of the *Mantis* database and the author of *The Smaller Majority*.

JAMES L. PATTON is Professor Emeritus in the Department of Integrative Biology and Curator in the Museum of Vertebrate Zoology at the University of California, Berkeley. His research interests focus on the biogeography of Amazonian mammals, and he is also a main contributor to the Grinnell Resurvey project in Yosemite National Park.

JOHN D. PERRINE is an Assistant Professor of Biology and the Associate Curator of Mammals at California Polytechnic State University in San Luis Obispo, California. His research and teaching focus on conservation biology and wildlife ecology. His current projects include the Grinnell Resurvey of mammals in the Lassen region of northern California, and the conservation and ecology of the Sierra Nevada red fox.

JAMES L. REVEAL, Adjunct Professor at Cornell University and Professor Emeritus at the University of Maryland, is an expert on botanical nomenclature, the family Polygonaceae, and the history of botanical exploration and discovery. In addition to writing over four hundred and fifty scientific papers, he is co-author of *Lewis and Clark's Green World*.

GEORGE B. SCHALLER's pioneering studies of endangered and little-known animals including gorillas, lions, giant pandas, as well as the wildlife of the Tibetan plateau, have provided scientific information critical to numerous conservation efforts. In addition, he has published many scientific papers and books, including *The Mountain Gorilla*, *The Serengeti Lion*, and *Stones of Silence*. Continuing his decades of leadership in field science and conservation, he currently serves at the Wildlife Conservation Society and as vice president of Panthera.

EDWARD O. WILSON is Pellegrino University Professor, Emeritus, at Harvard University. In addition to two Pulitzer Prizes (one of which he shares with Bert Hölldobler), Wilson has won many scientific awards, including the National Medal of Science and the Crafoord Prize of the Royal Swedish Academy of Sciences.

INDEX